Methodologies, Tools and New Developments for E-Learning

Methodologies, Tools and New Developments for E-Learning

Editor

Fernand Quincy

Methodologies, Tools and New Developments for E-Learning

Edited by **Fernand Quincy**

ISBN: 978-1-68117-232-3
Library of Congress Control Number: 2016934764

Preface

E-learning is essentially the network-enabled transfer of skills and knowledge. Educational technology is the effective use of technological tools in learning. As a concept, it concerns an array of tools, such as media, machines and networking hardware, as well as considering underlying theoretical perspectives for their effective application. Educational technology is not restricted to high technology. Also called e-learning, it includes an array of approaches, components, and delivery methods. E-learning is learning utilizing electronic technologies to access educational curriculum outside of a traditional classroom. In most cases, it refers to a course, program or degree delivered completely online. There are many terms used to describe learning that is delivered online, via the internet, ranging from Distance Education, to computerized electronic learning, online learning, internet learning and many others. E-learning is defined as courses that are specifically delivered via the internet to somewhere other than the classroom where the professor is teaching. It is not a course delivered via a DVD or CD-ROM, video tape or over a television channel. E-learning has definite benefits over traditional classroom training. While the most obvious are the flexibility and the cost savings from not having to travel or spend excess time away from work. With the resources provided by communication technologies, E-learning has been in employment in various universities, along with in wide range of training centers and schools. This book entitled Methodologies, Tools and New Developments for E-Learning covers the chapters dealing with the subject and stressing the importance of E-learning. It demonstrates the evolution of E-learning, with discussion about tools, methodologies, improvements and new opportunities for long-distance learning.

Table of Contents

Chapter 1   Design, Use and Evaluation of E-Learning Platforms:
             Experiences and Perspectives of a Practitioner from the
             Developing World Studying in the Developed
             World   1

Chapter 2   Effective eLearning and eTeaching — A Theoretical
             Model   23

Chapter 3   Scoping the Possibilities: Student Preferences towards
             Open Textbooks Adoption for E-Learning   59

Chapter 4   From Workshop to E-Learning: Using Technology-
             Enhanced "Intermediate Concept Measures" As a
             Framework for Pharmacy Ethics Education and
             Assessment   87

Chapter 5   Implementation of E-Learning System: Findings and
             Lessons Learned   111

Chapter 6   Efficacy and Usability of an E-Learning Program for
             Fostering Qualified Disease Management Nurses   127

Chapter 7   A Knowledge-building Process in Interaction-based E-
             Learning   147

Chapter 8   3D Interactions between Virtual Worlds and Real Life in
             an E-Learning Community   171

Chapter 9   Study of the Assessment Criteria on e-Learning
              Websites   197

Chapter 10   Considerations on Barriers to Effective E-learning toward
              Accessible Virtual Campuses   215

Chapter 11   Development of an Asynchronous Web Based E-Learning
              System   247

Chapter 12   Personalization and User Modeling in Adaptive E-
              Learning Systems for Schools   273

              Index   303

CHAPTER 1

Design, Use and Evaluation of E-Learning Platforms: Experiences and Perspectives of a Practitioner from the Developing World Studying in the Developed World

Nesba Yaa Anima Adzobu

Main Library, University of Cape Coast, University Post Office PMB, Cape Coast, GHA, Ghana; Tel.: +233-547-323-838

ABSTRACT

Electronic learning platforms are evolving and their evaluation is becoming more complex and challenging with time. Yet, the evaluation of electronic learning services is intrinsically linked to improving the performance of documentation services. In this paper, I describe my perspectives on the design, use and evaluation of an electronic learning platform using a lens of a practitioner from a third world country. I further delineate the challenges and constraints I encountered as a student learning about e-learning platforms and using e-learning platform services at an institution of higher learning in Sweden. In particular, the Ping Pong system at the University of Boras, Sweden, and the electronic print in the Library and Information Science (E-LIS), one of the services from the bulletin board for libraries

(BUBL) Link information gateway, will be evaluated. It is anticipated that this experiential evaluation will provide designers of e-learning platforms with insights and strategies for refining the e-learning platform to facilitate teaching activities and promote students' learning efficiency and satisfaction.

KEYWORDS

E-LIS; Ping Pong; user-centered; services; documentation; Europe

1. INTRODUCTION

Online learning platforms have come to stay; it is anticipated that they will co-evolve with traditional learning platforms in the future. Increasingly, e-learning platforms are taking center stage in learning and research in institutions of higher learning in developing countries. In this paper, I will attempt to examine the concepts of design, use and evaluation of e-learning platforms as explicated in the extant literature. I will also attempt to describe my personal experiences and perspectives on the design, use and evaluation of e-learning platforms regarding the untapped strengths of e-learning platforms in Ghana. In addition, I will enumerate the challenges and constraints I encountered as a practitioner from the third world (Ghana), learning about e-learning platforms and using e-learning platforms services at an institution of higher learning in a first world country (Sweden).

The design of e-learning platforms is a complex initiative. The complexity in design is reflected in the number of stakeholders involved in the various stages and processes of the design. This, indeed, calls for collaboration and interactivity at different levels of the design process. While collaboration on the one hand largely involves inter-stakeholder communication, interactivity on the other hand may require communication between humans and the system, a domain known as Human–Computer Interaction (HCI) [1]. This distinction is debatable because communication runs through both collaboration and interactivity. Interactive design is the design of a product or system that is helpful in allowing people to perform their daily task or work to meet usability goals and user experience goals [2]. "Usability goals" refer to the use of interactive products that are effective, efficient, memorable, learnable and safe, and have fewer errors from the user's perspective [2]. "User experience goals" refer to the feelings that cover all senses and are dependent on the user's prior experiences and values. It is the assessment of a product and the demand for its follow up. It can be argued that usability goals and user experience goals determine whether the design of a particular e-learning platform can be considered as satisfactory or unsatisfactory.

The goal of this paper is to assess the strengths and weaknesses of different kinds of functions offered by an e-learning platform in a developed country from the perspective of user from a third world country. For this experiential evaluation, I selected Ping Pong as my research target, the main e-learning platform used by the University of Boras in Sweden. Few evaluations of e-learning platforms exist in the extant literature regarding the African context. However, given that e-learning platforms are many and are increasingly playing prominent roles in altering the existing ways of teaching and learning in higher educational institutions in Africa, it is imperative to evaluate them in order to ascertain which of them best suits specific academic contexts. It is expected that this experiential evaluation will provide designers of e-learning platforms with insights and strategies for refining e-learning platforms to facilitate teaching activities and promote students' learning efficiency and satisfaction.

2. DESIGN OF E-LEARNING PLATFORMS

Let me now relate the various usability goals to my experience of the Ping Pong system (an e-learning platform for distance learners) at the University of Boras, Sweden. First, I received an e-mail containing a link to the Ping Pong site and I was instructed to log on to it for all my course materials including lecture notes, film lectures, downloads of course materials, assignments, submission of study task and study questions, communication between teachers, course mates and the systems co-coordinator. Ping Pong is a learning platform. Therefore, its primary purpose was to allow me to access course information, and to interact with other course mates and teachers in order to enhance teaching and learning. In relation to Ping Pong, all course materials and literature for the various units were provided on schedule. The deadline for submitting each task was provided. Links to course video lectures and indicative readings were also provided. All the information I needed to help me accomplish my task was provided and I had access to it. This suggests the system was effective. Effectiveness is one example of usability goals. According to Sharp *et al.* [2] "Effectiveness refers to how good [sic] a system is at accomplishing its intended purpose". However, the effectiveness could be much more complex than this. Hauck and Weisband [3] mention that finding ways to wade through the vast amount of information in large data stores in e-learning platforms is critical to using the system effectively. Yet, Toms [4] asserts that the fact that just because a system delivers what has been requested of it does not necessarily mean that the results meet the user's needs and are able to satisfy the requirements of the task. Throughout the course, every task demanded that I did an extensive and painstaking search before obtaining results that met the requirements of the task. Therefore, I think my experience with the Ping Pong system confirms the findings of Toms [4].

I encountered difficulties when using Ping Pong. The challenges I encountered were partly due to my limited previous knowledge about and the use of the system (in the past, I had not practiced any extensive use of such learning platforms; therefore Ping Pong was totally new to me). It was also partly due to lapses in the architecture of the Ping Pong system. In addition, Nielsen [5] identifies individual characteristics and differences and the user's task as two critical elements of usability. According Hauck and Weisband [3] in many cases, users who are experts with the technology may be the ones who decide if a new application or user interface is effective or not. In my case, sometimes the Ping Pong page either failed completely to load or it took a relatively long period of time to load. I do not need to be an expert in technology to know that this situation is bad. When this happened, my first reaction was to view my time code account to see how much money I was losing as a result of the delay in loading the Ping Pong page. At other times clicking on a menu led to a page where only scripts (some series of numbers and certain alphabets) appeared, a situation which indicated that the page had failed to load properly. Over the entire duration of the course, however, my common impression of Ping Pong was that it is generally effective. In Ping Pong, the course video lectures catered for high and low bandwidth users. I fall into the second category since I live in Sub-Saharan Africa, where information technology is still nascent. From the beginning of the first course, I was not able to access any of the course video lectures because the speed of the internet in my office was extremely slow. Of course, this problem was not caused by the Ping Pong system. The Ping Pong system did not allow me to download the video lectures. The message "cannot find link to cache" popped up whenever I attempted to download the video lectures, thus preventing me from accessing this necessary information. This situation reflects difficulties in the use of e-learning services. On this basis alone, Ping Pong may not be regarded as effective. The Ping Pong system design for the course video lectures should have been done in a manner that allowed students who currently live in Sub-Saharan Africa to download the course video lectures for offline viewing given that their internet speed could be extremely slow. My perspective on this matter is that I may be alone when it comes to the issue of inability to listen to video lectures. Access to relevant course information is critical to understanding the course better. Therefore, my inability to access this kind of information in one way or the other limited my capacity to deliver high quality assignments to teachers. Now, the critical questions I asked myself are: are the video lectures a question of bad designs? Would I have encountered this same problem if the video lectures were rather delivered in Audio or PDF format? Would all other students have appreciated audio or PDF lectures instead of video lectures, given the level of technological advancements today and the advantages of using video as against audio versions? Is it worth it to change the lecture format for the sake of only one student from the global south? What is the guarantee that the audio version of the lectures will even work for me? These questions are pertinent and useful for e-learning platforms designers because it will guide the design process and thus lead to better decision outcomes.

Apart from effectiveness, efficiency is another element of usability goals. According to Sharp et al. [2], efficiency refers to the help the system or product offers to its users. Basically, different people prefer or need different information architecture and different purposes require specific architecture.

Given that users of e-learning platforms come from diverse academic backgrounds and social persuasions, it implies that e-learning platforms need to be structured to allow all kinds of users to obtain access to information that might be used to meet their information needs. New generation students demand faster service as most campuses today are wired, allowing students to access e-learning resources from their laptops in their dormitories or in the classroom [6]. With regards to the Ping Pong system, I was provided with a user guide to assist me to explore the system. It helped me to understand the different menu displayed on the start page in relation to where and how to navigate to the course materials, where to view new messages, and how to use the right language in reading. The user guide also assisted me in understanding how to provide feedback when necessary, and how to communicate with the unit teacher by mail to seek clarification on questions and issues I did not understand. On the basis of these issues, Ping Pong could be regarded as efficient. The link to Ping Pong in itself was very easy to remember, very short and precise (http://pingpong.hb.se). My user name and password was also simple and short. These factors allowed me to easily login to the system. There was no need to write my user name and password down in my diary and carry it along every time and everywhere. When I needed to exit the system, the log out button is also placed at the far right hand corner of the system and is duly labeled; recently visited events are displayed on the start page which served as a reminder of the events I last visited. It also served as a short cut whenever I wanted to visit that same place. After signing in with my user name and password I first searched for the log out button, I had no problem tracing it as it situated at the same extreme right as in Yahoo mail. While I searched for the log out button in Ping Pong, I tapped this previous knowledge (experience) of where the log out button is found in Yahoo mail, and this helped me to easily locate the log out button. All the items in the Ping Pong start page were clearly labeled. Hence, even though I sometimes did not use the system for a week whenever I return to it I still recalled my previous actions. This enabled me to use Ping Pong often without difficulty. However, sending mails to multiple recipients in Ping Pong was a task that I failed to accomplish. In fact, I always had to copy the emails of recipients from elsewhere and paste it in the recipient address column before sending. This is a cumbersome process in terms of the time and energy spent in sending mails in Ping Pong. The help function in the system has not been of help regarding this matter. On this basis, the efficiency of Ping Pong could be questioned. It is argued that certain features of e-learning platforms (the more technical aspects) may discourage individuals from adopting a useful system since it may prove too difficult to use [3]. However, on a balance I think Ping Pong is generally efficient. Here, it is important to point out that my evaluation of Ping Pong as a system is subjective. It is based on the issue that I consider most important when it comes to efficiency. Another student may perceive this issue differently.

Learnability is another important issue when it comes to e-learning platforms design. Learnability of Ping Pong refers to the ease with which the guide to the Ping Pong system provides step by step guide to the system within a short time frame [2]. On the start page, menus such as how to use Communication, Personal, Events, Tools, Calendar collectively made the system easy to learn. On the whole, the Ping Pong system was easy for me to learn within a short period of time. Ping Pong was safe to use in that although my user ID and my password to the system was created beforehand and given to me, I exclusively had access to the password to my page and especially access to my documents. Teachers also had access to certain parts of the system when it came to download of submitted study tasks and study questions. All my colleagues were also given different user IDs and passwords. The password and the user identity were provided as privacy right for each student. Another safe use of Ping Pong is that the buttons are few and straight forward. There was no need for me to click many buttons in search of where to log out, or where to go the previous page or the start page.

Ping Pong provided the right kind of functionality so that I could make use of the system. Under the personal button, the system provided me with personal information on where to put my personal details, upload my photograph, and setting of my preferences (choosing how I wanted Ping Pong to appear). Other functionality included language settings and short cut routes. There was no time limit and I could make changes to my personal settings in the system whenever necessary. Ping Pong supports creativity, thus I added my personal details to the system. I was also able to create a new calendar and enter a new event on it, as and when it was necessary. Therefore Ping Pong shows a high degree of utility. Use of Ping Pong was motivating, satisfying and helpful because I received lecture notes on time, and I had access to all the information I needed in order to facilitate the learning process. Ping Pong was not necessarily entertaining because there are no games in the system. This was not a limitation of the system given that the primary purpose of Ping Pong is to educate. However, the use of Ping Pong was fun because I learnt new things on every visit and added to my previous knowledge every time I used the system.

As an inexperienced and new user of Ping Pong, some of the instructions, especially on how to submit my assignment, were not initially clear to me. This led me to submit my first two assignments wrongfully by placing them in my personal documents folder instead of loading it just below the study task. At the very beginning, I did not have enough time to go through the tour guide. I used the start page and clicked every item that displays a button or symbol for viewing its content. My initial aim was just to have an overview of the system and to be able to have access to my course material, read any messages available and to be able to send a Ping Pong Instant Message (PIM). With time, I did not click all the icons on the start page as I used to do when I started using Ping Pong initially. I visited the Ping Pong system almost every day. I subsequently got to know where to locate my personal documents, how to save my documents and how to send a PIM. Eventually, I got to know how to load my study task. I navigated easily and rapidly to all content of the system than when I started. This means that my user experience of the Ping Pong system immensely improved with time and frequency of use. Ping Pong is therefore efficient, effective, learnable, memorable, and safe and exhibits a high degree of utility. This notwithstanding, I think the system is far from being perfect. From the abovementioned usability goals, and from my subjective evaluation it appears that generally Ping Pong is a system that was able to perform its intended purpose and hence achieved its specific goals.

3. USE OF E-LEARNING PLATFORMS

According to Tammoro [7] and Monopoli *et al.* [8], users are satisfied when the speed of access to digital resources is high, access to materials is easy, only simple steps are required to find information and the availability of the materials is within reach. When users are offered access to a database of internet resources which they can search by keyword or browse by subject area they can do this in the knowledge that they are looking at a quality-controlled collection of resources [9]. When accessing information, users behave differently based on the environment (physical or virtual) within which the search is being carried out [10,11,12]. A number of theories have been proposed to explain the different behaviors that information users' exhibit in their quest to access information [13,14,15]. In this section, I will attempt to recount how I searched for information during my study- and work-related tasks. I will also attempt to relate theoretical explanations to my access of information. I observed that when I was performing study- and work-related tasks four behavioral patterns emerged. These behavioral patterns were interaction with metadata and real information, cherry picking, browsing and information encountering. These issues will be discussed in the order in which it is listed.

According to Pharo [11], searchers and users of e-learning platforms mainly interact with two kinds of information: bibliographic information (metadata) and the contents of materials (pure or real information). This statement is true of my search for information in both study- and work-related tasks. In study related tasks, I interacted with the following metadata: journal information such as title, volume, issue number and date, among others. In my study related tasks, this type of metadata serves as a kind of bridge to the information contained in each article because the title either kindles or dampens my interest. When the title aroused my interest I proceeded to read the abstract after which I downloaded the full article (real information). It is important to note that sometimes gathering metadata was not my fundamental goal or value when I was performing study-related tasks. On such occasions, metadata only served as a mechanism to reach the contents of each article. My goal or end for that reason is the information contained in each article. In performing my study-related tasks, metadata may be regarded as a means to an end. Hence, metadata is purely instrumental in this case. However, in carrying out work-related tasks, metadata is usually the goal.

One of my foremost work-related schedules at the University of Cape Coast School of Business library is cataloguing. The Library of Congress Online Catalogue is the database that I mainly use in performing work-related tasks. The basic search interface of this database provides search by the following metadata: title, author, subject, International Standard Book Number (ISBN), and Library of Congress Control Number (LCCN), among others. The specific metadata that I am interested in when I am carrying out work-related tasks is the call mark or class number. Therefore, the call number is my goal. However, title, author, subject, ISBN, and LCCN serve as the means to reach this goal that is the call number. Hence, in work-related tasks, a particular metadata serves as the goal while other metadata serve as the means to reach this goal. The steps involved in reaching my goal are as follows: typing the author's full into the Library of Congress Online Catalogue, clicking on a selection from a host of author names, the fitting author name that corresponds to the specific book being catalogued. Next, I carry out further searches by clicking the title of the book for bibliographic details. I then copy the call number or class mark except the author cutter number into the book that is being catalogued. If the author name is not found in the database, I shift from basic search to advanced search mode. If the advanced search mode fails to produce the call number or class mark, the book is put aside and I begin a new search using a different book. This behavior of mine in the performance of work-related tasks particularly in relation to author searching reflects the classic model of information retrieval as mentioned in Bates [16].

Cherry picking is also a behavior that characterizes some stages of my study- and work-related tasks. In my work-related task, I search for information for students. One particular example involved searching for information on the differences between a cell phone and personal computer, in a Management Information Systems course. In this case, I used Google search engine to gather general information from different sources without sorting it out or filtering it. All the information I gathered on the subject was transferred directly to a storage device from which the students had to sort out in a bid to meet their information need. During the search process, I realized that the search started on a broader note and moved through a series of sources from which I gathered bits and pieces of information. Along the line, querying kept changing in the search process. This kind of development in the search process reflects cherry picking [13,16]. Cherry picking is also reflected in my study-related search tasks. However, in this case I do the sorting and filtering of the information gathered, personally. According Bates [13] the system does not deliver a complete, single and final retrieved set. This makes it imperative to sort and filter the bits of information to meet my information need. Usually, I search for information outside the literature list provided in the course materials. During such search processes cherry picking is also evident particularly in the search of journal articles.

Another information searching behavior that appears in my study- and work-related tasks is browsing. As Bates [13] notes "browsing involves a series of glimpses, some glimpses leading to further, closer examination of things glimpsed and some not". This is a further elaboration on the model of browsing proposed by Rice *et al.* [17] which essentially considers it as a scanning activity or process. Interestingly, traces of the conceptions of browsing as proposed by Bates [13] and Rice *et al.* [17] became evident while I was performing my study- and work-related tasks. However, I realized that I predominantly used browsing as proposed by Rice *et al.* [17], at the beginning of the search process. The Bates [13] model of browsing became my predominant behavior as the search process progressed. I realized that the duration and speed of my browsing hinged on whether the information in the literature was relevant or not. Relevant information engaged my attention while I scanned through irrelevant information rapidly. Cothey [18] mentions that browsing strategies are iterative and are contingent on making out relevant information.

Information encountering is yet another behavior that was reflected in my work- and study-related tasks. According to Erdelez [10], this phenomenon appears to occur with "the unexpected discovery of useful and interesting information". Erdelez [10] argues that information encountering is much more than merely 'bumping into information'. Erdelez [10] further argues that even before the phenomenon occurs, the information searcher is positioned in one way or the other to receive this new information. In my case, I was consciously looking for information in relation to my study- and work-related tasks. While the search process was on-going, I saw information that might be interesting to my co-workers, friends and students. This suggests that I was in a sort of ready mode to receive information. It is significant to note that prior to my search, these individuals had already given indication of their information needs to me. In such situations, I proceeded to download the information for them.

Users will always be confronted by complexities and challenges when using e-learning platforms. The levels of complexities and challenges that the use of e-learning platforms brings will even be higher for students like me who come from the third world, where limited knowledge about and lack of access to new technology prevails. Speaking from a third world perspective, I think use of e-learning platforms in our part of the world is negligible, at best it is unsatisfactory. I am not sure whether most library users in the University of Cape Coast (both staff and students) are even aware of the full range of the prospects of e-learning platforms. On the awareness-use continuum, I can speculatively project that level of awareness and extent of use of e-learning platforms on campus among staff and students is low. In fact, until recently, the library had no staff with a postgraduate degree specifically in e-learning platforms. With the scant allocation of resources to fund research and establishment of e-learning platforms, limited technical expertise in the field, and over concentration on physical libraries in the university, I can definitely perceive the widening of the digital divide between the global North and South.

4. EVALUATION AND CONTEXT OF E-LEARNING PLATFORMS

Evaluation of all systems, not least an information system, is central to its sustainability. Evaluation is basically linked to information retrieval metrics such as precision, recall and fallout, among others. No doubt, it would be difficult for users to decide beforehand exactly what features they do and do not want, and to some extent the development of electronic devices is governed by an attempt on the part of the designers to "try it and see if there's a use for it." I have selected a digital service, namely electronic print in Library and Information Science (E-LIS), one of the services from bulletin board for libraries (BUBL) Link information gateway, as the object to evaluate. My motivation for choosing E-LIS is that I use this service regularly. I will consider the possibilities of evaluating E-LIS from a user's perspective. The evaluation of E-LIS will be based on goals, user groups (types of user communities), use of evaluation results and the instruments (methods) that would best suit the different goals and evaluation tasks.

Evaluation of services of e-learning platforms may take different forms or approaches [19,20,21,22,23]. This suggests that evaluating e-learning platforms is also a complex and difficult undertaking. However, according to Saracevic [22] systems approach is the predominant or widely used mechanism when it comes to evaluation of all information systems including e-learning platforms. A systems approach implies that e-learning platform services may be regarded as a system of interacting parts that function in concert to achieve specific targets. An evaluation can therefore be carried out on parts of or on the entire system to determine its effectiveness or efficiency or both [22]. Evaluation of the whole system could pose significant challenges to the evaluator compared with one that focuses on one or two parts of the system. Central to critical evaluation is the expertise needed to assess the system. The evaluator should usually have indices or values against which the system would be assessed. These could be objective or subjective.

Effectiveness and efficiency relate to performance. Therefore evaluation is primarily concerned with the performance of systems. Bollen and Luce [19], note that "the evaluation of services provided by e-learning platforms and collections is a multi-faceted problem that cuts across a wide range of systems, interfaces, and user communities as well as a multitude of issues in Human–Computer Interaction". They further contend that "any evaluation of Digital Library collections and services must inevitably take into account the characteristics of the Digital Library's user community". Marchionini, and Plaisant and Komlodi [24] share this contention. In this connection, Saracevic [22] identifies a number of evaluation types. One type of evaluation deals with user studies involving different user communities (students, teachers, researchers). The center of attention for this kind of evaluation is different design features in terms of usability and functionality with a view to improving design for the different user communities. Again, user logs have been used to establish user interactions through the interface with a view to developing better usability and functionality. Another type of evaluation has sought to use the instrument of interviews to understand different users, their socio-cultural settings, relative interests, their capacities to act (agency), their opportunities, their constraints and their goals. In effect, this kind of evaluation explores how individuals use of tools and technologies within specific socio-contextual settings. Yet another type of evaluation explores how different users identify, retrieve, read and use materials in articles of interests.

From the above, it appears that each approach to evaluation serves a different a purpose. Each approach has merits and demerits. Therefore no single approach is superior to the others. In fact, it is the goal of the evaluation that determines whether the approach is appropriate or not. Saracevic [22] further identifies a number of approaches to evaluation. These include ethnographic which is well suited to the attainment of a general understanding of the function and outcome of a practice or a construct in a wider collective or group framework. Sociological approach is appropriate for shedding light on social forces and effects. Economic approach is apt when it comes to accounting for economic factors such as investment cost, return on investment and payback time. It may be considered as a kind of cost-benefit analysis to determine whether continued financial resource allocation to a project on e-learning platforms is justified or not. A situation in which economic analysis was instrumental in financial decision-making is reflected in Choudhury, Hobbs and Lorie [25]. A political approach concentrates on policy and political factors. For instance, what kind of institutional structures in terms of policy and the legislative framework must be put in place to facilitate the efficient running of an e-learning platform's service? These issues and many more shape the evaluation of e-learning platforms.

5. DESCRIPTION OF THE E-LIS

E-LIS is an archive for materials in library and information sciences, which was formed in 2003. It is the first international e-server in this subject area. E-LIS relies on the voluntary work of individuals from diverse backgrounds and is non-commercial. The purpose of the E-LIS library archive is to make full text LIS documents visible, accessible, harvestable, searchable and usable by any potential user with access to the internet. The materials available in the archive include books, book chapters, journal articles, conference proceedings, conference posters, conference papers, thesis, working papers, newspapers, magazines, bibliographies, manuals, tutorials and instructional materials, among others. The wide array of materials available in the archive reflects the scope of the archive in meeting the needs of various users. Metadata for most of the materials is available. These include abstract, additional information, alternative locations, author names, conference dates and locations, country, department and Editors. The contents of the archive are accessible by search (quick, simple, advanced) and browse (by year, subject, authors/editors, books/journals, country). In terms of browsing by subject, E-LIS uses the JITA classification system. JITA is an acronym for the first names of Jose Manuel Barrueco Cruz, Imma Subirats Coll, Thomas Krichel and Antonella De Robbio. JITA was developed after a merger of the News Agent Topic Classification Scheme (maintained by Mike Keen at Aberystwyth, UK, until 31 March 1998) and the RIS classification scheme of the (now defunct) Review of Information Science, originally conceived by Donald Soergel (University of Maryland). Searching and archiving in E-LIS are free for any user.

5.1. Evaluation of the E-LIS

5.1.1. EVALUATION GOALS OF THE E-LIS

Basically, evaluation goals of E-LIS can focus on either the system or the user. In considering goals of E-LIS as a system, the processing, engineering and content of E-LIS are the focal points. Here, by examining feedback from users of E-LIS through for example log analysis, the system designers can develop new technologies that support a range of search strategies from hierarchical selections to formal and comprehensive queries so that the needs of beginners and experts are both met. This is geared towards improved performance of the system. Evaluation goals of E-LIS may include better learning for all user groups. It may take account of improved research for specific user groups. Improved dissemination and communication between user and system may also feature as a possible evaluation goal for E-LIS. This particular goal suggests collaboration. E-LIS can also be evaluated based on usage patterns-time of visit or use of the service, preferences of users, duration of search, and the response time for feedback. Bollen and Luce [19] outline a methodology for generation of such usage patterns using electronic learning server logs. This approach is useful when it comes to improvements of e-learning platform collection organization.

E-LIS can also be evaluated based on economic goals such as cost-benefit analysis. Cost-benefit analysis takes into account the qualitative value of e-learning platform collections and services to users. Even if digital libraries had a clear definition of what it means to be cost-effective or a benchmark against which to measure their cost-effectiveness, additional work is required to determine whether the benefits of an activity warrant the costs. If the cost of an activity is high and the payback is low, the activity may be revised or abandoned. In the same vein, cost-benefit analysis can be carried out on E-LIS to justify its continued existence. In considering goals of E-LIS from the perspective of the user, it will be appropriate to capture the user from different social levels-users as individuals, as institutions, and as society or communities. This introduces an element of complexity or diversity into the evaluation. Users of E-LIS include content providers, students, teachers and researchers. Within each category of user group and across the groups there are variations in their specific needs and the technological settings within which they work. For instance Marchionini, Plaisant and Komlodi [24] argue that users exhibit a wide array of individual characteristics, preferences, and experiences. An undergraduate student may not have the same preferences or experiences as a postgraduate student even though both are students. Researchers may not necessarily have identical backgrounds (historians, chemists, sociologists, anthropologists, authors, *etc.*). E-LIS users may also show this kind of diversity therefore capturing such user taxonomies in evaluation of E-LIS is crucial to guiding E-LIS interface design.

It is possible to construct user taxonomies based on motivation, domain knowledge, E-LIS system knowledge, focus, and time allocations as was carried out by Marchionini, Plaisant and Komlodi [24]. The usefulness of such taxonomies in evaluation resides in the continued development of e-learning platform interfaces. It may also be useful when it comes to meeting specific needs (queries) and providing contextual information (scope of need). The methods used to construct the taxonomies may include distribution of questionnaire to user communities in E-LIS and examination of the documents in E-LIS with a view to unearthing interface challenges in terms of content and users and strategies. Eventually the results regardless of the instrument used can be employed to improve user satisfaction with the existing E-LIS service.

5.1.2. THE USE OF RESULTS AND METHODS OF EVALUATION THAT CAN BE USED FOR SPECIFIC GOALS

A number of methods or instruments for evaluating library services have been identified in the literature. Generally these instruments are either qualitative or quantitative. They include surveys (questionnaires), focus groups, user protocols and transaction log analysis, among others. Surveys are usually quantitative and but can sometimes be qualitative. It is very useful in gathering information about E-LIS users' previous or current behaviors, attitudes, beliefs and feelings. The results of the survey may be used by the E-LIS editorial board to drive different directions of strategic planning. For E-LIS system designers, the results of surveys may assist them to improve service quality in terms of reducing response time. It may also help them in setting priorities; inform customization and E-LIS website vocabulary revision. Focus groups discussions are exploratory, guided interactions among seven to ten participants with common interests. The common interest in this case refers to use of the E-LIS library service. Insights from focus group discussions can inform resource allocation and strategic planning. For instance, the administrator of E-LIS can use the results to identify user problems and preferences related to E-LIS collections format and thereby increase user satisfaction. User protocols may be employed to gather in-depth insight into the behavior and experience of a person using the E-LIS tools. This is instrumental in the identification of problems in the design, functionality, navigation and vocabulary of the E-LIS website. The results can assist E-LIS system designers in rearranging the ELIS hierarchy, changing the order and presentation of search results in E-LIS and revising the metadata classification scheme for text collections in E-LIS.

A Transaction Log Analysis can basically be used to evaluate E-LIS system performance. It may be employed to study unobtrusively interactions between users and the E-LIS website. It may also be employed to track patterns of use by different user communities and the distribution of use across communities. The results can be used to construct usage patterns over time, understand user needs and inform interface redesign.

6. CONCLUSIONS

The design, use and evaluation of e-learning platforms reveal complexities and challenges. The interaction of users with e-learning platforms shows that the process is not smooth or unproblematic. This is particularly so for individuals whose backgrounds (both professional and social) reflect limited previous use and understanding of such services. In fact, the difficulties manifest at different levels of the design, use and evaluation process. It is unlikely that the challenges associated with using e-learning platforms will disappear soon. Since the design and use of e-learning platforms is an iterative process, it implies that the process is evolving, and every new day brings its own difficulties that need to be overcome. The difficulties may arise from different aspects (interface, architecture, navigation, *etc.*) of the design. This calls for collaboration and inclusion of diverse stakeholders in a bid to finding lasting solutions to such complexities and difficulties associated with the design, use and evaluation of e-learning platforms. The design, use and evaluation of e-learning platforms is a vast field. On their own, each of them is an extensive field. Therefore, this paper cannot claim to have comprehensively dealt with all the questions surrounding the three fields in a satisfactory manner. Not only do I lack the skills, experience and expertise to undertake such a huge and almost impossible task, but also space constraints will not permit it. The aim is to give a brief and broad overview of the main issues. In terms of the scope of the topic dealing with the design, use and evaluation of e-learning platforms, the commentary in this paper is just the tip of the iceberg.

REFERENCES

1. Dix, A. *Human-Computer Interaction*; Springer: New York, NY, USA, 2009; pp. 1327–1331.

2. Sharp, H.; Rogers, Y.; Preece, J. *Interaction Design: Beyond Human-Computer Interaction*, 2nd ed.; John Wiley and Sons: New York, NY, USA, 2011; Chapters 1 and 4.

3. Hauck, R.V.; Weisband, S. When a better interface and easy navigation aren't enough: Examining the information architecture in a law enforcement agency. *J. Am. Soc. Inf. Sci. Technol.* **2002**, *53*, 846–854.

4. Toms, E. Information interaction: Providing a framework for information architecture. *J. Am. Soc. Inf. Sci. Technol.* **2002**, *53*, 855–862.

5. Nielsen, J. *Usability Engineering*; Morgan Kaufmann: San Francisco, CA, USA, 1993.

6. Lukasiewicz, A. Exploring the roles of academic libraries. *Libr. Rev.* **2007**, *56*, 821–827.

7.   Tammaro, A.M. User perceptions of digital libraries. *Perform. Meas. Metr.* **2008**, *9*, 130–137.

8.   Monopoli, M.; Nicholas, D.; Panagiotis, G.; Korfiati, M. A user-oriented evaluation of digital libraries. *Aslib Proc.* **2002**,*54*, 103–117.

9.   Monopoli, M.; Nicholas, D. A user-centered approach to the evaluation of Subject Based Information Gateways: Case study ADAM. *Aslib Proc.* **2001**, *53*, 39–52.

10. Erdelez, S. Information encountering: It's more than just bumping into information. *Bull. Am. Soc. Inf. Sci. Technol.***1999**, *25*, 26–29.

11. Pharo, N. A New Model of Information Behaviour based on Search Situation Transition Schema. *Inf. Res.* 2004, 10. Available online: http://InformationR.net/ir/10-1/paper203.html (accessed on 28 April 2014).

12. Smith, A.G. Search features of digital libraries. *Inf. Res.* 2000, 5. Available online: http://informationr.net/ir/5-3/paper73.html (accessed on 25 March 2014).

13. Bates, M. What is browsing—really? A model drawing from behavioural science research. *Inf. Res.* 2007, 12. Available online: http://informationr.net/ir/12-4/paper330.html (accessed on 10 January 2014).

14. GodBold, N. Beyond information seeking: Towards a general model of information behaviour. *Inf. Res.* 2006, 11. Available online: http://InformationR.net/ir/11-4/paper269.html (accessed on 10th January 2014).

15. Wilson, T.D. Models in information behaviour research. *J. Doc.* 1999, 55, pp. 249–270. Available online: http://InformationR.net/ir/9-1/paper164.html (accessed on 25 January 2014).

16. Bates, M.J. The design of browsing and berry picking techniques for online search interface. *Online Inf. Rev.* **1989**, *13*, 407–424.

17. Rice, R.E.; McCreadie, M.; Chang, S.-J.L. *Accessing and Browsing Information and Communication*; MIT Press: Cambridge, MA, USA, 2001; Chapters 3 and 9.

18. Cothey, V. A longitudinal Study of World Wide Web Users' Information Searching Behaviour. *J. Am. Soc. Inf. Sci. Technol.* **2002**, *53*, 67–78.

19. Bollen, J.; Luce, R. Evaluation of digital library impact and user communities by analysis of usage patterns. *D-Lib Mag.* **2002**, *8*, 1–13.

20. Covey, D.T. *Usage and Usability Assessment: Practices and Concerns*; Digital Library Federation Council on Library and Information Resources: Washington, DC, USA, 2002.

21. Marchionini, G. *Evaluating Digital Libraries: A longitudinal and Multifaceted View*, Draft Version ed; Graduate School of Library and Information Science, University of Illinois at Urbana-Champaign: Urbana/Champaign, IL, USA, 2000.

22. Saracevic, T. Digital library evaluation: Toward an evolution of concepts. *Libr. Trends* **2000**, *49*, 350–369.

23. Salampasis, M.; Diamantaras, M. Experimental User-Centered Evaluation of an Open Hypermedia System and Web Information Seeking Environments. *J. Digit. Inf.* 2002, 2. or alternatively: http://www.webcitation.org/5bn17aN1B.

24. Marchionini, G.; Plaisant, C.; Komlodi, A. The people in digital libraries: Multifaceted approaches to assessing needs and impact. In *Digital Library Use: Social Practice in Design and Evaluation*; Bishop, A.P., van House, N.A., Buttenfield, B.P., Eds.; MIT Press: Cambridge, MA, USA, 2003; Chapter 6.

25. Choudhury, S.; Hobbs, B.; Lorie, M.; Flores, N. A framework for evaluating digital library services. *D-Lib Mag.* **2002**, *8*, 1082–9873.

CHAPTER 2

Effective eLearning and eTeaching — A Theoretical Model

Maureen Snow Andrade[1]

[1] *Academic Programs, Office of Academic Affairs Utah Valley University, Orem, UT, USA*

ABSTRACT

Distance learning is increasingly becoming an option for learners that were previously denied higher education opportunities due to elitist systems, cost, academic preparation, or personal circumstances. It is also a means to help nations meet goals to increase the percentage of individuals with post-secondary education in order to address workforce needs. However, learners and instructors often have concerns with their ability to be successful in a distance learning environment. This chapter presents a theoretical model for eLearning and eTeaching aimed at helping learners and instructors successfully navigate distance learning courses. Examples of course activities corresponding to the model components are shared. A qualitative analysis of learner self-reflections demonstrates the efficacy of the model in terms of increased autonomy, self-regulation, and targeted skills.

KEYWORDS

distance language learning, self-regulation, eTeaching and eLearning, online instructor training, distance education

1. INTRODUCTION

Knowledge, an end in itself as well as the surest route to higher wages and longer lives, is measured by degree attainment and school enrollment [1]. eLearning provides access to higher education for a wide range of learners. These include traditional students in university classrooms, individuals in the workplace seeking to formalize their work experience through the pursuit of a degree or to transition into a different career path, and those who need flexible scheduling options or prefer to not participate in traditional learning environments. These learners can select courses and degree programs from institutions of higher education in their local areas or from providers across the globe. eLearning increases access, which has historically been denied to many due to elite education systems, and offers choice—choice in providers, programs, scheduling, cost, and content.

Elitist views of tertiary education are receding [2-6], with recognition of the benefits of a well-educated workforce in terms of economic development, economic stability, health and well-being, and decreased crime [7], factors that are encouraging governments to lower the barriers to higher education and set goals for degree attainment [7,8]. The appeal of eLearning is readily recognized as instrumental in these endeavors [8-11], evidenced by increased enrollments in online courses. In the United States, for example, the number of university students taking an online course increased from 1.6 million to 7.1 million in a 10-year time span (2002-2012) [12]. Many of these learners are nontraditional in terms of age, marital condition, and employment status [13].

In spite of demand, involving university instructors in developing, implementing, and teaching online courses can be challenging due to concerns with quality, nontraditional methods of interaction with students, low student performance, pedagogical skill, technological knowledge, workload, time intensiveness, large class sizes, and course ownership [14-16]. Learners may struggle with the discipline needed in eLearning contexts, which are generally less structured than face-to-face settings, feel intimidated by the technological expertise required to navigate courses and submit assignments, lack motivation, or simply be convinced that learning in a more traditional format in which they can engage in class with other learners and an instructor is preferable.

This chapter introduces common challenges with eLearning in terms of learner success and instructor expertise, and suggests solutions to these challenges through the framework of self-regulated learning [17-21] and the supporting theories of transactional distance [22-24] and collaborative control [25]. The chapter provides a guiding model for course design and pedagogy, illustrated with specific course content and activities. The theoretical model of eLearning and eTeaching helps learners overcome barriers to success while parallel training based on the same principles prepares instructors to facilitate effective online learning experiences. Learner and teacher self-reflections were examined to identify the presence of the model's components, and are shared to demonstrate the model's efficacy. Additional recommendations for evaluating the model are provided. The approach is illustrated with online English language courses and a related instructor training course; however, the model and its elements can be applied to courses in any discipline and be examined quantitatively or qualitatively to determine its effectiveness in facilitating learner success.

2. PROBLEM STATEMENT

While global growth in technology-based learning, and online learning in particular, presents significant opportunities for learners to access higher levels of education that were previously out of reach, and for institutions wanting to address challenges associated with the resources needed to expand their physical infrastructure to accommodate enrollment increases, or desiring to take advantage of outreach beyond their state or national borders, the fact remains that many stakeholders are concerned with the efficacy of this delivery method. Anecdotes of negative distance teaching and learning experiences abound on many university campuses in spite of growing evidence to the contrary and increasing interest and participation. Views toward distance education tend to be polarizing, as expressed in the following statement:

At one end of this continuum, we detect what some might argue is an overly sanguine view of what distance education has already achieved and how much it has influenced pedagogy and the academy. At the other extreme is the pessimistic perspective that this phenomenon is a scourge threatening the quality and integrity of academe [26].

Related to the latter extreme, concerns involve accreditation, institutional support, scalability, technological literacy, instructional strategies, rigor, expertise, and fear of taking missteps into this new territory due to possible negative repercussions. Each of these issues has and can be addressed. The focus of this chapter is on online teaching and learning, and specifically, approaches that situate both learners and instructors to have a positive experience.

3. THEORETICAL FOUNDATIONS: LITERATURE REVIEW

In all educational contexts, every effort must be made to ensure that learners succeed. This involves pedagogical considerations, understanding learner backgrounds and approaches to learning, instructor skill, and course design and management. Distance learning presents its own set of factors related to success such as a less structured experience for learners in that they do not meet regularly in a classroom; a possible learning curve related to course delivery technologies, which is a potential concern for both learners and instructors; different strategies for sharing and discussing information than would be present in a face-to-face context; and specific to teachers, the need to adapt and expand on traditional face-to-face instructional tools. In effect, novice online learners and instructors must be prepared and supported in this new learning endeavor. While success for students in any learning context, and particularly in distance learning, is dependent on a number of factors, not all of which are within the control of an instructor, much can be done to anticipate and alleviate challenges inherent in an online course. Similarly, instructors who have a solid understanding of online teaching approaches and the ability to apply them will be able to provide a more positive learning environment for their students and fully enjoy their teaching experience.

Self-regulated learning [17-21] is an educational theory which can be maximized in an online class to provide learners with the needed scaffolding to manage their learning. The theory of transactional distance [22-24], from the field of distance education, provides insights into the relationship among the course, learners, and instructor, and how the psychological distance created by the physical gap between the learner and teacher can be mitigated. The concept of collaborative control [25], most frequently applied to distance language learning, addresses the myth that online learning is synonymous with independent learning (although this is a possibility), and suggests strategies for collaboration. These three concepts can be applied to course design and instructor training to maximize the opportunities associated with distance learning and assist learners and instructors in developing the requisite skills and abilities for success.

4. SELF-REGULATED LEARNING

The concept of self-regulated learning has been applied to the teaching and learning process to increase student achievement across age, educational levels, and delivery modes [17,18, 27-33]. It is most commonly defined as the ability to control the elements and circumstances that affect learning [17,18]. A useful framework is the six-dimension model, which consists of motive, method, time, social environment, physical environment, and performance [17,18, 19-21]. These dimensions address the questions why, how, when, with whom, where, and what. Figure 1 provides additional details about the dimensions.

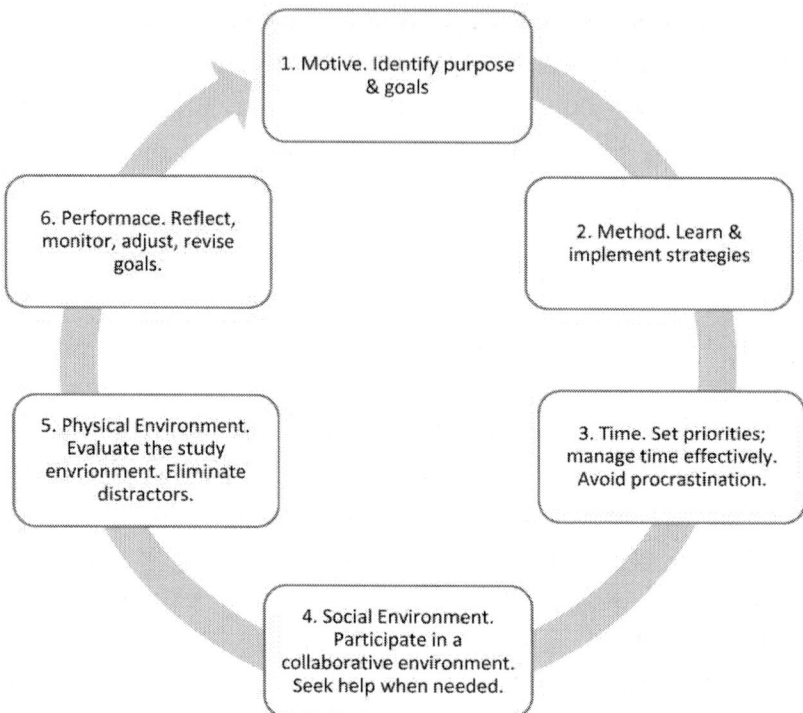

Figure 1. The cycle of self-regulated learning.

Another commonly applied self-regulated learning framework consists of three phases—"the *forethought* phase refers to processes and beliefs that occur *before* efforts to learn, the *performance* phase refers to processes that occur *during* behavioral implementation, and *self-reflection* refers to processes that occur *after* each learning effort" [21, p. 67]. Forethought is similar to *motive* in that it encompasses goal setting, motivation, and examining beliefs about learning. Performance is most closely related to *methods* as depicted Figure 1; it involves strategy identification, implementation, and monitoring of the strategies applied to determine their effectiveness. Methods might also include consideration of areas such as time management and social and physical environments in that learners must determine strategies to improve their performance by using time effectively, getting help from others, and eliminating physical distractors where and when they study. Finally, self-reflection consists of reviewing learning outcomes and comparing them to a standard. It is similar to the performance dimension in Figure 1, which involves examining the achievement of goals and learning outcomes and determining the next steps.

While both frameworks are valuable, the former is particularly helpful in guiding learners. The specificity of the various areas needing consideration assists learners in analyzing their current practices in relation to their study approaches and making needed changes by following the cyclical process of goal-setting, strategy implementation, control of time and environmental factors, and review of performance. Goals can be set related to any of the dimensions (e.g., methods, time, social environment, physical environment) depending on individual strengths and weaknesses. Methods involve strategies for processing and acquiring knowledge and improving skill. In the case of language learning, strategies might focus on vocabulary learning with note cards, using the context to guess the meaning of words, recognizing error patterns in written grammatical usage, or using headings to find the main ideas in reading passages. Outside of language study, more general learning methods can be applied such as various strategies for reading (e.g., previewing, mapping main ideas and supporting details, asking questions, paraphrasing, and summarizing), studying and recalling information (e.g., listening, identifying transitional phrases, note-taking), writing (organization, idea development, revision, and editing), improving performance on assignments and tests (e.g., following instructions, understanding distractors in multiple choice tests, organizing a response to a short answer or essay exam question), and interaction and discussion (teamwork, collaboration, class participation). Many methods or learning or strategies are applicable across content areas.

The six-dimension framework can serve as the foundation to online course design to assist learners in controlling the factors that affect their learning concurrent with gaining content knowledge for a given discipline or increasing skills to perform specific tasks. Course assignments involving goal setting, examining motive for learning, gaining awareness of various learning strategies, recognizing the role of the social and physical environments in learning, and monitoring and reflecting on performance can be integrated into the course. This design approach can increase the achievement of learning outcomes [27,29], particularly if the self-regulated learning components are required rather than optional; students rarely complete optional assignments. For many students, considering various approaches to learning is completely new, as illustrated in the following quotation from an English language learner who discusses his use of methods of learning, specifically, taking reading notes and composing study questions.

Taking this course helps me to know how important it is to use my study materials. Throughout my junior and high school, I always thought that using my study materials was a waste of time because I thought that I wouldn't really understand if I used my study materials compared to asking my teachers to explain how to do the activity. Every time I read an assignment my mind was not really focused. I didn't pay attention so I ended up not knowing what the assignment was all about. I didn't make notes of what I read or write down questions as I read the assignment or paragraph. But today I can say that using my study materials is really important.

This quote provides evidence that the learner has become more autonomous through the use of methods of learning. Instead of relying on the teacher to explain assignments, he has recognized that he can read, take notes, and write down questions to help him understand the material. This indicates self-regulation.

5. TRANSACTIONAL DISTANCE AND COLLABORATIVE CONTROL

Two other theories have relevance to online course design and instructor support and can be integrated with the self-regulated learning framework. The first is the theory of transactional distance [22-24]. This theory has three components: structure, dialogue, and autonomy. Structure is provided by course materials, content, assignments, and deadlines, which are fixed prior to a course being made available to students. Structure is a helpful pedagogical tool as it provides learners with predictability in determining how the course is organized, sequencing of instructional modules and assignments, and deadlines for assignment submission. Generally, courses are designed so that each lesson has the same sections and various pages in the course have the same formatting, similar to a textbook. This helps learners know what to expect as they progress through the course. A course syllabus is also part of the structure as it guides students through the course and its requirements and provides needed information about policies, procedures, requirements, and grading. Structure can also be provided through communication for the purpose of guiding and supporting learners. This communication is referred to as dialogue, and includes any type of exchange or interaction in a course. It can be among the learners or between learners and the instructor. It includes course features such as discussion boards, peer review of assignments, instructor announcements, instructor feedback on assignments, student questions, e-mail, and live conferencing. Dialogue provides socialization, particularly through peer interaction. Instructor dialogue can motivate learners, help them identify their strengths and weaknesses, and assist them in making needed improvements.

The amount of structure and dialogue in a course affects autonomy, which is defined as choice characterized by elements of self-direction. Lower levels of structure and dialogue support greater levels of autonomy. Some learners are able to function well with low levels of structure and dialogue while others need greater support. Autonomous learners are able to determine learning goals and steps for reaching those goals. They have both instrumental independence and emotional independence [22-24]. In other words, they can progress through the course with little help and need little encouragement. One could expect that learners might become more autonomous over the weeks they are enrolled in an online course as they understand expectations and gain confidence in their ability to be successful. Instructors can facilitate this confidence building. The goal of structure, dialogue, and autonomy is to support the achievement of learning outcomes and prevent student attrition. The theory of transactional distance has similarities to that of self-regulated learning. The self-regulated learning framework inFigure 1 provides a type of structure for learners to help them set goals, practice strategy application, make use of dialogue (i.e., the social environment) to get help when needed, and self-evaluate in order to have greater capacity for autonomous learning.

Finally, the concept of collaborative control [25] provides greater understanding of the social environment aspect of self-regulated learning and the dialogue component of the theory of transactional distance. Rather than conceiving of distance learning as an independent activity, collaborative control acknowledges that learners can learn from and help each other and also that the instructor can facilitate learner interaction and success. Help-seeking is a positive practice as long as learners are not overly dependent on others. They need to recognize when they need help, identify the best sources of help, and evaluate the effectiveness of the help received [34]. As the name of the concept implies, collaborative control occurs when learners and the instructor collaborate to complete tasks, thereby improving learning outcomes. Instructors must be aware that the goal of collaboration is to encourage greater levels of self-regulation or autonomy so that learners can make sound choices and have the confidence to succeed. However, as a common criticism of distance learning is the purported lack of social interaction and exchange among learners; thus, course designers and instructors should always be aware of opportunities to provide for this aspect of learning.

6. THEORETICAL INTEGRATION

Figure 2 demonstrates the integration of various aspects of the three theories. Online courses that require learners to engage in forethought (goal-setting), performance (strategy application and monitoring), and self-reflection (review of progress); that are designed with a specific structure (organized content), opportunities for dialogue (peer and instructor communications), use of the social environment (help-seeking), and collaborative control (learner and teacher collaboration on tasks), all of which are facilitated by the instructor; and that use these features to help learners gradually develop greater autonomy (capacity for self-direction and making choices), self-regulation (ability to control factors affecting learning), and targeted skills and knowledge (course content and related outcomes) as they set goals, apply what they are learning, and reflect on their learning demonstrate how the three theories work synergistically to improve the online learning experience.

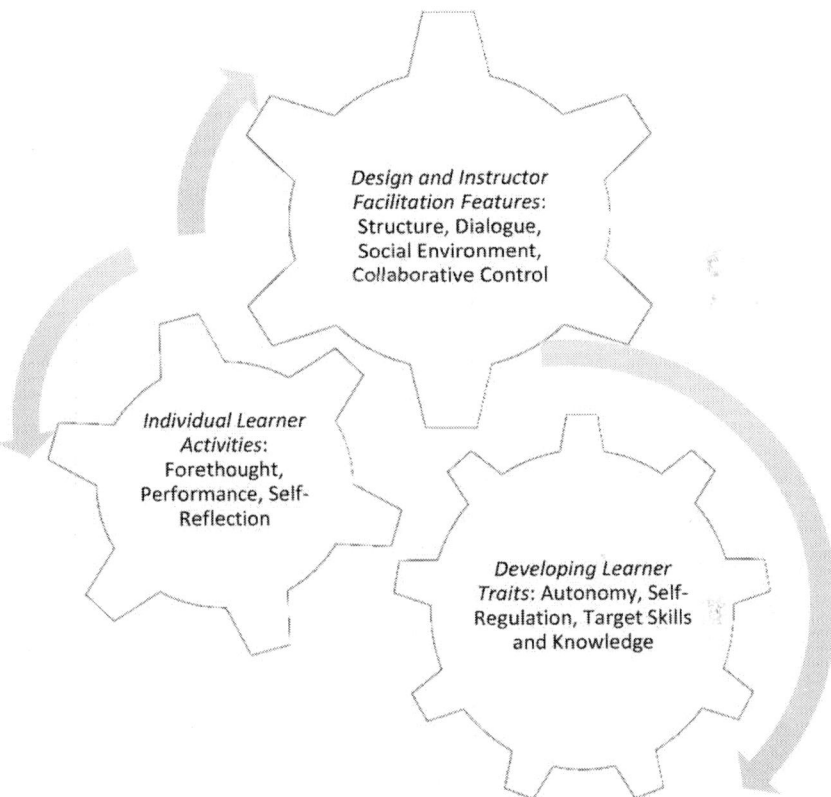

Figure 2. Model of eLearning and eTeaching.

The model demonstrates how the theories of self-regulation and transactional distance and the practice of collaborative control can guide distance learning and teaching approaches. Although the design and instructor facilitation features in the model overlap, each concept has distinct characteristics that help to inform design choices and instructional approaches. The model can be applied to both online student courses and related training for online instructors. In the latter case, the learner is the instructor.

7. APPLYING THE MODEL: PRINCIPLES AND PRACTICE

To address the challenges inherent in eLearning, course designers and instructors must consider ways to facilitate effective student learning and course completion. In the process, students enrolled in the course can acquire lifelong learning strategies with broad applicability. Additionally, administrators, or those responsible for assigning instructors to courses, must ensure that these instructors are prepared for and skilled in online teaching. This can be accomplished through required instructor training that models the student online learning experience.

This section of the chapter identifies principles based on the theoretical model presented in Figure 2. The principles are designed to help learners be successful eLearners and to prepare instructors to make the transition from traditional pedagogies to those that facilitate eLearning [27,30]. A primary consideration for instructors is communication with learners through dialogue and response strategies [22-24]. The discussion is organized around the three areas of the model of eLearning and eTeaching (see Figure 2). For each area, key principles are identified with examples of applications for a student course and an associated teacher training course.

8. INDIVIDUAL LEARNER ACTIVITIES

While every online course contains a variety of learner activities, the focus in this section is on those related to the three phases of self-regulated learning: forethought, performance, and self-reflection. For purposes of delineating this portion of the model, both students and instructors are considered learners; both are applying the principles of self-regulated learning in their respective courses—students are learning about a specific content area or discipline or acquiring a related skill while instructors are honing their online pedagogical skills. The courses provide the opportunity for individuals in each group to set goals, apply strategies, and reflect on their performance. An additional benefit of these self-regulated learning activities for instructors is that they are experiencing what their students will be doing in the student course. It is strongly recommended that these activities be required so that all learners will benefit from them.

The forethought stage of self-regulated learning can be integrated into online courses and online teacher training through activities that help individuals examine their beliefs about learning, diagnose their strengths and weaknesses in the subject area or in relation to needed teaching and learning strategies, and to set goals [28, 29, 31]. The six-component framework in Figure 1 can help learners generate areas of focus for goals as can some type of diagnostic evaluation. The forethought stage should be initiated at the beginning of the course, possibly in the introductory module of a course. It helps set expectations for learner responsibility and raise awareness of course content.

Next, the performance stage involves the use and monitoring of strategies. Strategies should be introduced and linked to the assignments in the course. For example, if students are required to compose a writing assignment, part of the instruction for the assignment might include strategies for revision and editing or learning how to evaluate the appropriateness of sources [34]. It could also involve discussion of the particular genre expected, e.g., a critique of an art exhibit for an art class or the summary of an academic journal article for a psychology course. In keeping with the six dimensions of self-regulated learning [17-21], strategies could also be introduced related to effective time management or evaluating the physical environment to determine its conduciveness to learning. In the teacher training course, strategies would include those related to online pedagogies such as using whole-class feedback rather than responding to students individually, providing supplementary materials to address identified learner weaknesses, facilitating a discussion board, or incorporating the use of a new form of technology [27, 29, 30].

The third and last area, self-reflection, is designed to help learners evaluate the benefits of the various strategies they have applied. They can then examine reasons for their outcomes and modify their methods accordingly. "Overall, the available research evidence suggests that promoting self-reflection, self-regulation and self-monitoring leads to more positive online learning outcomes. Features such as prompts for reflection, self-explanation and self-monitoring strategies have shown promise for improving online learning outcomes" [35, p. 45]. As such, this is critical for students and also important for teachers as it involves taking the time to carefully consider teaching approaches and also to experience directly what they are asking their students to do. Possible activities for each of the three areas are summarized in Table 1.

TABLE 1. Activities for the Three Phases of Self-regulated Learning

Principle Learning can be improved through identifying the purpose for learning, goal-setting, and examining beliefs about learning (forethought); learning, practicing, and monitoring strategies (performance); measuring performance against a self-imposed or external benchmark, and modifying goals and strategies as needed (self-reflection).	Student Course	Instructor Training Course
Forethought	· Introduction activity in which learners post information about themselves, their	· Introduction activity in which teachers post information on a discussion board about

background in the subject area, and their purpose for learning; posts can be written or oral (video recordings); learners are required to respond to a specific number of peer posts.
· Appoint students to take turns to facilitate the discussion board throughout the semester to increase comprehension of the subject matter and help them gain confidence.
· Introductory writing assignment stating previous experience with the subject area and reasons for wanting to learn more. Peer or teacher response to assignment.
· Completion of an instrument identifying learners' self-beliefs about learning (e.g., intelligence is fixed vs. intelligence can be developed) [36].
· Diagnostic survey to help learners identify strengths and weaknesses related to the subject area or to the use of academic learning strategies.
· Identification of goals related to strengths and

themselves, their teaching or professional backgrounds, and their purpose for teaching online; posts can be written or oral (video recordings); teachers are required to respond to a specific number of peer posts.
· Appoint teachers to take turns to facilitate the discussion board throughout the semester to practice the skills they will use in the course they are teaching.
· Introductory writing assignment stating previous experience with online learning and reasons for wanting to learn more. Peer or trainer response to assignment.
· Writing assignment may include asking teachers to provide their philosophy of teaching, perspectives of online teaching and learning, previous successes with distance education as a student or a teacher, or any other prompt that

	weaknesses in the subject area or learning in general; goals should be specific, measurable, achievable, result-oriented, and time-bound (SMART).	helps teachers dialogue with each other and create community. · Online survey listing various online teaching practices, particularly those relevant to the course, from which teachers determine their strengths and weaknesses. · Identification of goals related to what teachers hope to learn in the course and what they anticipate their needs to be in terms of online teaching based on the survey.
Performance	· Strategy instruction integrated with course content; learners apply strategies as they complete course assignments. · Activity choices focused on the six dimensions of self-regulated learning (motive, methods, time, physical environment, social environment, performance). · Opportunities to select appropriate strategies to accomplish course assignments.	· Learn about and practice methods for online instruction and responding to students. This instruction includes becoming familiar with the theoretical foundation for the course. · Opportunities to adapt familiar face-to-face teaching strategies to an online environment. · Introduction to technology-based instruction and

		application activities, including tools available through the course management system.
Self-reflection	· Discussion boards, learning journals, or survey instruments that provide prompts for reflection on goals at regular intervals in the course (e.g., weekly, monthly, midterm, end of course). · Opportunities to share reflections with peers or the instructor for feedback. · Inclusion of goal modification and next steps as part of reflection.	· Discussion boards, learning journals, or survey instruments that provide prompts for reflection on goals at regular intervals in the course (e.g., weekly, monthly, midterm, end of course). · Opportunities to share reflections with colleagues or the trainer for feedback. · Inclusion of goal modification and next steps as part of reflection.

9. DESIGN AND INSTRUCTOR FACILITATION FEATURES

Although learners engage in the three phases of self-regulated learning independently, and set goals and practice the specific dimensions of self-regulated learning (motive, method, time, physical environment, social environment, performance) largely on their own, the process is facilitated through course design and instructor dialogue. The four components of this part of the model—structure, dialogue, social environment, and collaborative control—have a theoretical basis, described in the previous section, and are built into the course design. They also have implications for instructor behavior.

As indicated earlier, the structure of the course helps guide learners and provides predictability while dialogue entails communication among course participants and the teacher for purposes of socialization and learning support. These two elements affect autonomy. Figure 3 indicates the relationship among structure, dialogue, and autonomy.

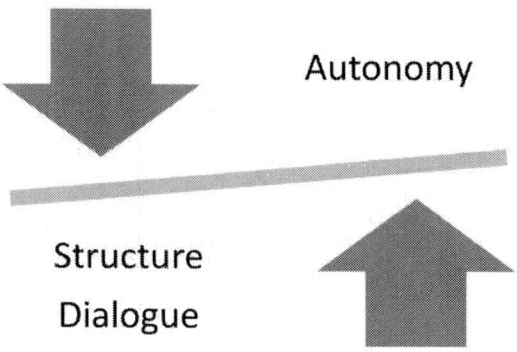

Figure 3. The interaction among structure, dialogue, and autonomy.

When structure and dialogue are low, the learner is able to make choices about learning independently. However, structure and dialogue may be needed when necessary information is not included in the course or if the information is incomplete, or when learners need greater levels of support. When structure and dialogue are high, autonomy decreases. As instructors work with students, they can tailor the instruction and support to the level of the learner through the use of dialogue. As the learners' skills improve, the amount of dialogue will likely decrease. The following example demonstrates structure in the form of assignment instructions that are set in the course, and instructor dialogue that provides additional information to help learners successfully compete the assignment. Instructor dialogue can be provided as an announcement (oral or written), a supplemental page in the course, in a brief video, or in an e-mail to the class.

9.1. Assignment

Read the article about the spread of English as an international language and then write a one-paragraph summary of the main ideas. Organize your ideas clearly and upload the paragraph for your teacher's review.

9.2. Instructor Dialogue

In order to complete this assignment, you need to know how to write a summary. See the steps below to help you complete the article summary assignment.

1.   Understanding the reading

The first step in writing a summary is to read through the article carefully and make sure you understand it.

- Check any vocabulary you don't know

- Find the author's thesis statement or main point

- Find the main idea (topic sentence) of each paragraph or section of the article

- Underline main ideas as you read

- Reread the article and take notes—use your own words; put exact words in quotation marks

- Explain what you read to someone else

2.   Writing the first sentence

The first sentence of a summary follows a specific format. See the instructions and examples that follow.

- The first sentence of a summary includes the following:

    o   The author's name—usually the family name only

    o   The title of the article, book, magazine, or chapter

    o   A restatement of the author's thesis

    o   Examples:

- According to their article, "A Model for Self-regulated Distance Language Learning," Andrade and Bunker claim that learning a language through distance education is a challenge due to the learner's limited opportunity for real-life interaction.

- In "A Model for Self-regulated Distance Language Learning," Andrade and Bunker observe that distance language learning is particularly difficult due to the lack of opportunity for learners to hear and use the language.

3.   Completing the summary

When you are certain you understand the content of the reading and have practiced expressing it orally, and you have written your first sentence for the summary using one of the patterns, use your notes to draft the remainder of the summary paragraph. Be sure to review the paragraph carefully and edit as needed so that your language is as accurate as possible.

- Write a sentence or two summarizing each main idea of section of thought

- Put the sentences in the same order of the ideas in the original article

- Use transitions to connect the ideas

- Check your use of punctuation

- Check sentence structure, verb tense, and other grammar

- Add your reference list (see the writing tip on how to format references)

- Remember that a summary

  o   explains the main ideas of the original article in your own words

  o   does not contain your own opinion

  o   is usually 1/4 to 1/3 the length of the original

The assignment instructions assume that learners know how to write a summary. The dialogue, in the form of supplemental materials, provided by the instructor gives them specific steps to follow. The teacher also needs to include an example summary paragraph and instruction about reference list formatting. Even though the materials in an online course are already established; in other words, the structure is present, skilled online teachers know their learners and recognize when additional information is needed. This situation is similar to using a textbook in a course; the teacher determines what to use from the textbook and what to add based on a needs analysis of the learner. As instructors teach a course regularly, they have a better sense of where learners will struggle and how to assist them. This is a principle that needs to be included in online teacher training and addresses the viewpoint that is sometimes present among instructors that online teachers are glorified graders. The opposite is true—effective pedagogy and interaction is critical in an online course.

The next example demonstrates how instructors can facilitate learning through dialogue in the form of response to assignments. The instructions for the assignment are an example of structure, complete with specifics for how to formulate the post, word count, and deadlines, while the instructor feedback guides learners toward better utilization of this structure so as to meet expectations for the assignment.

Assignment

Post a response to ONE of the following questions about the article you summarized. Clearly state your point of view and use information from the reading and your own experience to support your answer. Respond to at least two of your classmates' posts. Your initial post should be at least 100-150 words in length. Responses to your classmates must be 50-75 words. Your initial post is due Monday at 11:59 pm. Your two responses to classmates are due Thursday at 11:59 pm. Be sure to check the discussion each day.

Questions

1. Is English an instrument for global economic expansion or does it simply increase inequalities between those who are proficient and those who aren't? Explain.

2. Does the spread of English decrease cultural uniqueness across the world or does it serve to connect people across cultures and increase intercultural understanding? Explain.

3. Will English maintain its status as a global language or will another language supplant it at some point in the future? Explain.

Student post

English provides people with opportunities for education and better jobs. This is why I am studying English. I want to go to school in the UK. Then, when I come back home, I can get work and look after my future wife and children better.

Instructor dialogue

You have important reasons for wanting to improve your English skills. Please make sure you focus on the topic in the question. You have mostly talked about your personal reasons for wanting to learn English. You need to bring in ideas from the article, take a position on the topic, and support your viewpoint. These are skills that you will need in academic writing. The discussion board assignment is designed to help you practice and develop these writing skills. You also need to make sure your response is 100-150 words. Always make sure you read the instructions and follow them carefully. If there is something you don't understand, I am here to help. Please also review the example responses that I posted in the announcements this week. I look forward to your next post.

The instructor response example follows the sandwich approach, which entails starting and ending with positive comments (the two slices of bread) while the middle portion (the filling) provides specific recommendations for improvement. The purpose is to encourage learners while helping them improve. Instructor dialogue in the form of feedback on assignments can be individualized, as in the example above, or it can be provided to the entire class, which is the purpose of the announcement the instructor refers to. The whole-class approach allows the instructor to provide a more extensive response with examples and models. To do this, the instructor notes commonalities in strengths and weaknesses in the students' interactions and provides feedback, examples, and direction to the entire class similar to what would occur in a face-to-face course. Some of this information, depending on the length, could also be posted in the discussion board itself—during the time period that students are doing the assignment. Students do need to know where to find the information, so it is best to have a common practice for this.

Providing a rubric for assignments and including it with the instructions also provides structure and helps learners know how they will be graded. However, often learners need to be reminded to examine the rubric carefully and compare their work to it. An option related to this is to use the practice of collaborative control to help learners develop their own evaluation abilities. Learners could be paired or assigned to small groups and asked to review and discuss each other's posts and then evaluate them with the rubric, providing rationale for the scores assigned. Although students may tend to give each other high scores, the activity at least increases their awareness of the rubric and helps them explore common or differing understandings of it. This is also an example of how the social environment can be integrated into a course. Another way to use the social environment is to simply have an open discussion board in which students post their own questions about anything related to the course to other students. In this way, students can get help on their own that is not instructor-focused. This supports the development of autonomy and self-regulated learning in that they are selecting topics of need and taking responsibility for factors affecting their learning.

This discussion has included guidelines for instructors related to the components of structure, dialogue, social environment, and collaborative control. In the instructor course, structure is provided in the set course modules on various topics—introduction to the university mission and English language program, logistics about the course they will be teaching and the learning management system, understanding their role as online instructors, planning and preparing, and exploring techniques for response. Dialogue and the social environment are incorporated into the course design with opportunities for instructors to share perspectives on the information in the various modules such as how to apply the teaching strategies presented. Examples such as those shared regarding student posts and corresponding instructor feedback can be included in the course to help teachers see models of response. Additionally, strategies for individual and whole-class response, use of technology, and other pedagogical information relating to the principles of structure, dialogue, social environment, and collaborative control should be included and practiced in the training course. The instruction needs to introduce these features, the philosophical foundation for the student course and related research, and also give the teachers the opportunity to discuss these areas much in the same way the learners are discussing the global nature of English in their discussion board in the example shared previously.

10. DEVELOPING LEARNER TRAITS

The goal of online instruction based on the model of eLearning and eTeaching is characterized in following quote: "The capacity to learn autonomously is seen to develop from a state of self-awareness and willingness to take an active part. In order for learners to achieve this state, teachers must also play their part" [37, p. 220]. The first component of the model accounts for the need to develop learners' awareness and engagement by having them examine their purpose for learning, beliefs about learning, and by setting goals. They take an active part as they learn and apply strategies, monitor their outcomes, and modify their approaches. Taking an active part also involves the social environment and dialogue with others in the course. The model provides guidance for how instructors can "play their part" to help learners become more self-aware, fully participate in the learning process, and take control of their learning. Teachers facilitate this through practices related to structure, dialogue, the social environment, and collaborative control. They model these behaviors in the teacher training course.

The aim of the model of eLearning and eTeaching, reflected in the third component—developing learner traits—is to help learners develop greater capacity for autonomy and self-regulation, and to meet learning objectives related to the targeted skills and knowledge of the course. For teachers in a training course, the model assists in giving them direct experience with self-regulation and related theoretical components in order to understand the course design philosophy and how they can facilitate the goals of the model as they interact with learners.

In this section, I review possible methods for determining evidence of the model validity. The first possibility involves using the self-reflection instruments that are built into the course as a means to determine learners' experiences and their views of their own learning. In both the student and the teacher training courses, participants are involved in self-reflection through some type of journal, discussion board post, formal paper, or video recording. These may be required at various points in the course such as weekly or at the middle and end of the term. Prompts for reflection can be specific to the course content, focused on learning in general, and be formative or summative as the examples in Table 2 illustrate. All of the examples are focused on the self-reflection aspect of self-regulated learning.

11. RESEARCH ON MODEL VALIDITY

One method of providing evidence of the efficacy of the model is to examine learner self-reflections. The examples below, from students and teachers, illustrate learner perceptions of various aspects of the model, and particularly the six dimensions of self-regulated learning (motive, methods, time, physical environment, social environment, performance). The examples are taken from weekly, midterm, and final reflections in the student class, and end-of-module and end-of-course reflections for teachers. The model elements are indicated in parentheses following each quotation.

Table 2. Self-reflection Prompts and Theoretical Connections

Purpose	Reflection Prompt Students	Formative Summative	Reflection Prompt Teachers	Formative Summative	Model/Theory Elements
Specific to course content and activities	What did you learn about organizing a paragraph in this week's lesson? How did it make a difference in the way you write?	Summative	How can you help students become more self-regulated? What types of feedback can you provide that will help students increase their self-regulated behaviors and strategies?	Formative	Methods Dialogue Social environment
	What was the most helpful thing you learned about vocabulary study this week? How will you apply this in the future?	Formative	What did you learn this week about types of technology you can use to communicate with students? How will this make a difference in your teaching this semester?	Formative	Methods Dialogue Social environment
	How did the course help you improve your academic writing skills?	Summative	How did the training course help you change your perspective of online learning and teaching?	Summative	Methods Performance
	Review the diagnostic survey you took at the beginning of the semester and the goals you set based on the areas in which you wanted to improve. Evaluate your progress on your goals and explain what you will do in the next half of the course to continue to make improvement. This may include modifying your goals.	Formative	What strategies did you learn in the module for creating a community of learners online? Which of these will you use in your course? Why?	Formative	Forethought Performance Dialogue Social Environment
General learning/ teaching strategies	Think about the readings you discussed with your	Formative	What do you think of the whole-class	Formative	Forethought Performance

Purpose	Reflection Prompt Students	Formative Summative	Reflection Prompt Teachers	Formative Summative	Model/Theory Elements
	classmates this week. Share one idea or insight you learned from a classmate. How will that insight help you be a better learner?		feedback approach? How is it similar to what you would do in a face-to-face class? How do you think you might utilize it?		Dialogue Social environment Collaboeative control
	How have you used feedback from your tutor appointment to improve your language skills or solve a problem?	Summative	What can you do to provide feedback to students to help them synthesize their learning for the week, identify needed areas of improvement, and set related goals? How can you help students recognize the progress they are making?	Formative	Performance Methods Forethought Dialogue Social environment Collaboeative control
	How will you apply the learning strategies you practiced in this course to future learning? Give an example.	Formative	How do you view your role as an online teacher? How can you apply your skill as a face-to-face teacher to an online environment? What advantages and disadvantages are there to having the content of the course set? How can you respond to student needs when the content is already established in the course?	Formative	Methods Performance Dialogue
	Discuss the two most effective strategies or skills you learned in the course.	Summative	Think about your experience this semester as an online teacher. Consider the materials in the course and the goals you set for each unit. Reflect on how your perspectives have changed.	Summative	Performance Methods Forethought

Purpose	Reflection Prompt Students	Formative Summative	Reflection Prompt Teachers	Formative Summative	Model/Theory Elements
			Comment on your goals and to what degree you achieved them.		

Students

- I learned a lot of things through this wonderful activity about identifying values and setting goals. First of all, I need to understand my motivation before I set goals Sometimes, it was difficult for me to set a meaningful goal for I did not understand my desire and vision. I was like a visitor without a compass and map, and I got lost. Therefore, I need to evaluate my motivation and needs before I set my goals. (Motive)

- I reviewed the goals I set for the course to see if I am making progress. I need to see if I have made any changes. I need to make some changes from what I have done before to now in order to get better grades. These are new goals for the second half: first, try to distinguish between important things and unimportant things. Second, try to improve reading speed, and the last one is improve on finding main ideas in readings (Performance, Methods, Target Skills)

- The course activities are good activities because these activities can help me to become a better learner. Reviewing the activities, I found that I learned a lot. For example, I found out my learning styles, my strengths and weaknesses as an English language learner, and my reading strategies. These three activities help me to study better. Besides, I found out how to schedule my study time and have a study plan and to study in short segments. These two activities help me to have a better time management. (Methods, Time, Performance)

- Online distance learning program are quite strenuous compared to traditional classroom learning, notwithstanding this fact I felt it was fun and balanced. The structures set up to help the student know, do and become is just awesome, most significant of all is sharing ideas and learning from each other, furthermore our speaking partners made it more interesting. (Structure, Dialogue, Social Environment, Collaborative Control)

Instructors

- My goal is to create an assignment where students write an explanation of the directions BEFORE they do it. Then they need to write a self-evaluation on how well they followed the directions. My hope is to get a better idea of how much they understand from the directions, as well as for them to better understand and evaluate what is expected of them and how well they are meeting these expectations. (Forethought, Methods, Target Skills)

- The whole-class feedback approach is helpful. There are sometimes patterns of mistakes and collective misunderstandings that we should recognize and respond to. I will use this approach and I believe students can really benefit from it. It is my responsibility to give them extra tips, advice, or even reviews so they can make progress. (Forethought, Methods, Target Skills)

- I have been taking notes on students' individual weaknesses and strengths all semester. I can see how those who have followed my advice and have actively worked on their mistakes are actually improving faster than those who are going through the motions only. (Methods, Performance, Target Skills)

- I have tried individual and whole-class feedback. They both work well. The discussions have been one of the most effective ways for students to interact with each other. I would like to try video feedback and online office hours in the future. (Methods, Performance, Target Skills).

Compiling these comments and using qualitative analysis methods to determine categories and themes [38] can provide insights into the effectiveness of the course activities and learners' evaluations of their success in achieving course outcomes. The examples above indicate clear evidence that students and teachers benefitted from the structure and dialogue in the course, the opportunity to examine their motivation, goal setting activities, engagement in the social environment, and the instruction and practice activities related to methods and strategies. Outcomes indicate that they acquired new strategies, increased their self-regulation behaviors, and achieved the target skills and knowledge for the courses.

Another way to measure the effectiveness of the model is to have students complete a formal survey with specific questions about course activities and then analyze the data. Questions might be both forced choice (e.g., multiple choice or ranking) and open-ended. Forced choice questions could consist of asking learners to rank the activities in the course grouped by model component (e.g., for dialogue/social environment/collaborative control, these could include the discussion board, video postings, peer review of writing drafts, group writing project), evaluate the instructor on a Likert scale (e.g., value of instructor feedback, response time, knowledge, quality of interactions), or the effectiveness of the course design and content (ease of navigation, technology use, organization of the home page, clarity of instructions). Examples of open-ended questions might include the following: What aspects of the course did you feel were the most valuable? What specific learning (or teaching strategies) were new to you and which did you have the most success with? What were the greatest strengths of your instructor? What suggestions for improvement do you have for your instructor? This approach is a little more structured than the course self-reflections, particularly if quantitative responses are included. These can be collected over several semesters and the results compared to help inform administrators and designers of needed changes in the content and structure of the course.

Additionally, instructors should be given the opportunity to provide feedback about the course materials. As they are teaching, they will notice inconsistencies in the course, unclear instructions, portions of the course that students have difficulty with, and poorly worded or incorrect content and test items. They will also have suggestions for content changes and different pedagogical and presentation approaches. While course designers and administrators responsible for the course would not want to implement all of these recommendations or respond immediately except in cases where simple corrections or typos are needed, this feedback should be compiled and analyzed to determine needed revisions to the course. Also, if the course is part of a sequence of courses, feedback should be obtained from instructors teaching the next level course in the series to determine if the course and instructional techniques are preparing learners adequately.

Other ways to evaluate the model would be through course completion and test scores. The latter would be dependent on ensuring that instructors are rating student work and applying rubrics consistently in the case of assignments such as essays or projects. Comparisons could be made between courses in which teachers have been trained and those in which instructors have not been trained in the model or between courses with embedded learning strategy activities and those without.

12. IMPLICATIONS AND CONCLUSIONS

A limitation of the information presented in this chapter is that the model of eLearning and eTeaching has been applied to only English language learning courses and related teacher training courses, although the theoretical components of the model are well-established as being effective in improving learning in both online and face-to-face contexts [17-21]. Further application of the model should be extended to other disciplines and those using it should assess its value in helping learners and teachers become successful in an online context.

The model does much to address concerns with online learning and teaching from students and faculty members. It provides students with structure and the potential to improve their strategies and approaches to learning so as to be effective in a technology-based distance course. It addresses instructor concerns and myths about teaching online, particularly that online instructors are limited to a grading rather than a teaching role. It expands awareness on the part of both the learner and the teacher of the value of taking responsibility for learning and the role of autonomy, and addresses the misperception that distance courses involve largely independent study with no social interaction or learning from others. It also helps define the role of the instructor in an online course and indicates how instructors can facilitate autonomy through collaborative control, a concept expressed in the following quotation: "What matters most in language advising for autonomy, however, is the [teacher's] ability to help learners make informed decisions about their learning without making those decisions for them" [39,

p. 17]. This is the goal of instruction—to help learners develop the capacity to make sound decisions about what and how to learn.

Evidence of the efficacy of the model has been established through a qualitative examination of learner and teacher self-reflections that are embedded into both the student and teacher training courses. This examination demonstrated that the elements of self-regulated learning, transactional distance, and collaborative control, when applied in English language learning and teacher training contexts, assists participants in goal setting, the application of new learning and teaching methods, reflection on these methods, and improved performance. Practices and activities involving dialogue, the social environment, and collaborative control support this process and provide critical elements for the development of autonomy, the acquisition of targeted skills, and overall, successful eLearning and eTeaching.

With increasing demand for higher education, and movements in the United States, the United Kingdom, Europe, and elsewhere to expand the numbers of individuals with post-secondary education [2-4, 8,9], eLearning will continue to grow to fill this need. Indeed, the use of technology, and specifically, online learning, is a strategy to reduce the time spent in class and provide alternate pathways that support degree completion by allowing students greater access to the coursework they need. This can help students avoid excess credits, address the common problem that courses are not offered when needed, not offered at convenient times, or not offered frequently enough [40]. Online learning is convenient to the schedules and lifestyles of today's diverse learners in higher education [13]. However, those involved in its development and delivery must take action to ensure its success for all stakeholders. The model of eLearning and eTeaching is an important step in this direction.

REFERENCES

1. Lewis K, Burd-Sharps S. American Human Development Report. The Measure of America. Brooklyn, NY: American Human Development Project and the United Way; 2014. http:// www. measureofamerica. org/wp-content/uploads/2013/06/MOA-III.pdf (accessed 29 November 2014).

2. Corver M. Trends in Young Participation in Higher Education: Core Results for England. Bristol, UK: Higher Education Funding Council for England; 2010. http://www.hefce.ac.uk/pubs/hefce/2010/10_03/ (accessed 29 November 2014).

3. Higher Education Funding Council for England. Strategic Plan 2006-11; 2009. http://www.hefce.ac.uk/pubs/hefce/2009/09_21/ (accessed 29 November 2014).

4. Higher Education Funding Council for England. National Strategy for Access and Student Success in Higher Education; 2014. https://www.gov.uk/government/uploads/system/uploads/attachment_d ata/file/299689/bis-14-516-national-strategy-for-access-and-student-success.pdf (accessed 29 November 2014).

5. Trow MA. Reflections on the Transition from Elite to Mass to Universal Access: Forms and Phases of Higher Education in Modern Societies since WWII. Berkeley, CA: UC Berkeley Institute of Governmental Studies; 2005. http://escholarship.org/uc/item/96p3s213 (accessed 29 November 2014).

6. Kamenetz A. DIY U: Edupunks, Edupreneurs, and the Coming Transformation of Higher Education. White River Junction, VT: Chelsea Green Publishing Company; 2010.

7. Baum S, Ma J, Payea K. Education Pays 2013: The Benefits of Higher Education for Individuals and Society. Washington, DC: College Board; 2013. http://trends.collegeboard.org/sites/default/files/education-pays-2013-full-report.pdf (accessed 29 November 2014).

8. Lee JM, Rawls A. The College Completion Agenda, 2010 Progress Report. New York, NY: National College Board Advocacy and Policy Center; 2010. http://completionagenda. collegeboard.org/sites/ default/ files/reports_pdf/Progress_Report_2010.pdf (accessed 29 November 2014).

9. Lumina Foundation for Education. A Stronger Nation through Higher Education. Indianapolis, IN; 2010. http://www.luminafoundation.org/

states_landing/a_stronger_nation_through_education/ (accessed 29 November 2014).

10. International Council for Open and Distance Education, European Association of Distance Teaching Universities. Maastricht Message 2009. Oslo, Norway, Heerlen, Netherlands; 2009. http://www.ou.nl/Docs/Campagnes/ICDE2009/M-2009_Maastricht_Message.pdf (accessed 29 November 2014).

11. Altbach PG, Reisberg L, Rumbley LE. Trends in Global Higher Education: Tracking an Academic Revolution. Paris: UNESCO; 2010. http://www.uis.unesco.org/Library/Documents/trends-global-higher-education-2009-world-conference-en.pdf (accessed 29 November 2014).

12. Allen IE, Seaman J. Grade Change: Tracking Online Education in the United States. Oakland, CA: Babson Survey Research Group, San Francisco, CA: Quahog Research Group, LLC; 2014. http://www.onlinelearningsurvey.com/reports/gradechange.pdf (accessed 29 November 2014).

13. Radford AW. Learning at a Distance. Undergraduate Enrollment in Distance Education Courses and Degree Programs. Washington, DC: U.S. Department of Education; 2011. http://nces.ed.gov/ pubsearch/pubsinfo.asp?pubid=2012154 (accessed 29 November 2014).

14. Pundak D, Dvir Y. Engineering college lecturers reluctance to adopt online courses. European Journal of Open, Distance, and e-Learning 2014. DOI: 10.1080/01587919.2014.926803. http://www.eurodl.org/?p=current&article=637 (accessed 29 November 2014).

15. Shulte M. University instructors' perceptions of factors in distance education transactions. Online Journal of Distance Learning Administration: 2010;13(11). http://www.westga.edu/~ distance/ojdla/summer132/schulte132.html (accessed 29 November 2014).

16. Worthen H. How the Working Conditions of Online Teaching Affect the Work Lives of Online Faculty: Report from the Coalition of Contingent Academic Labor (COCAL) United Association for Labor Education (UALE) Working Group on Online Learning Survey, October-December 2012; 2013. http://uale.org/papers-and-presentations-from-2013-conference (accessed 29 November 2014).

17. Dembo MH, Eaton MJ. Self-regulation of academic learning in middle-level schools. Elementary School Journal 2000; 100(5) 473-490.

18. Dembo MH, Junge LG, Lynch R. Becoming a Self-regulated Learner: Implications for Web-Based Education. In: O'Neil HF, Perez, RS. (eds)

Web-Based Learning: Theory, Research, and Practice. Mahwah, NJ: Erlbaum; 2006. p. 473-490.

19. Zimmerman BJ. Self-regulated learning and academic achievement: an overview. Educational Psychologist 1990;25(1) 3-17.

20. Zimmerman BJ, Risemberg R. Self-regulatory dimensions of academic learning and motivation. In: Phye, GD (ed) Handbook of Academic Learning: Construction of Knowledge. San Diego, CA: Academic Press; 1997. p. 105-125.

21. Zimmerman BJ. Becoming a self-regulated learner. Theory Into Practice 2002:41(2) 64-70.

22. Moore, MG. Learner autonomy: The second dimension of independent learning. Convergence 1972;5(2): 76-88. http://www.ajde.com/ Documents/theory.pdf (accessed 15 January 2015).

23. Moore, MG. The theory of transactional distance. In: Moore MG. (ed.), Handbook of Distance Education (2nd ed.), Mahwah, NJ: Lawrence Erlbaum; 2007. p. 89-105.

24. Moore MG. The theory of transactional distance. In: Moore, MG (ed.), Handbook of Distance Education (3rd ed.). Mahwah, NJ: Lawrence Erlbaum; 2013. p. 66-85.

25. White C. Language Learning in Distance Education. Cambridge: Cambridge University Press; 2003.

26. Beaudoin, M. Issues in distance education: A primer for higher education decision-makers. In: Andrade MS (ed.) New Directions in Higher Education. Special Issue on Distance Education. Hoboken, NJ: John Wiley & Sons (in press).

27. Andrade MS. Building community and learner autonomy: An instructor training model for online learning. Middle East and North Africa (MENA) Higher Education Leadership Forum: Proceedings of the 2nd Middle East and North Africa (MENA) Higher Education Leadership Forum, 10-11 November 2015, Dubai, United Arab Emirates.

28. Andrade MS. Course embedded support for online English language learners. Open Praxis 2014;6(1): 65-73. http://openpraxis.org/ index.php/OpenPraxis/issue/view/7/showToc (accessed 29 November 2014).

29. Andrade MS. Dialogue and structure: Enabling learner self-regulation in technology enhanced learning environments. European Journal of Educational Research 2014;13(5): 563-574. www.wwwords.eu/eerj/ content/pdfs/13/issue13_5.asp (accessed 29 November 2014).

30. Andrade MS (2014, August). Effective whole class feedback for second language writers. Annual Conference on Distance Teaching and Learning: Proceedings of the 30th Annual Conference on Distance Teaching and Learning, 12-14 August 2014, Madison, Wisconsin.

31. Andrade MS. Self-regulated learning activities: supporting success in online courses. In: Moore, JS (ed.) Distance Learning. Rijeka: InTech; 2012. p. 111-132. Available from http://www.intechopen.com/articles/show/title/self-regulated-learning-activities-supporting-success-in-online-courses (accessed 29 November 2014).

32. Andrade MS, Bunker EL. Developing self-regulated distance language learners: A promising practice. Targeted Cooperative Network of European Institutions (STELLAR-TACONET): Proceedings of the Fourth Annual Self-regulated Learning in Technology Enhanced Learning Environments Conference, 10-12 May 2011, Barcelona, Spain.

33. Andrade MS, Bunker EL. Language learning from a distance: a new model for success. Distance Education 2009;30(1): 47-61.

34. Andrade, MS, Evans, NW. Principles and Practices for Teacher Response in Second Language Writing: Developing Self-regulated Learners. New York: Routledge; 2013.

35. Means, B, Toyama, Y, Murphy, R, Bakia, M, Jones, K. Evaluation of Evidence-Based Practices in Online Learning: A Meta-Analysis and Review of Online Learning Studies. Washington, DC: U.S. Department of Education; 2010. http://www2.ed.gov/rschstat/eval/tech/evidence-based-practices/finalreport.pdf (accessed 15 January 2015).

36. Means, B., Toyama, Y., Murphy, R., Bakia, M., & Jones, K. Evaluation of Evidence-Based Practices in Online Learning: A Meta-Analysis and Review of Online Learning Studies. Washington, DC: U.S. Department of Education; 2010. http://www2.ed.gov/rschstat/eval/tech/evidence-based-practices/finalreport.pdf (accessed 15 January 2015).

37. Hurd, S. Autonomy at any price? Issues and concerns from a British HE perspective. Foreign Language Annals 1998; 31(2), 219-230.

38. Strauss, A. & Corbin, J. Basics of Qualitative Research: Grounded Theory Procedures and Techniques, Newbury Park, CA: Sage Publications; 1990.

39. Benson, P. What's new in autonomy? The Language Teacher 2012; 35(4) 15-18.

40. Complete College America. Time is the Enemy. http://completecollege.org/docs/Time_Is_the_Enemy.pdf (accessed 29 November 2014).

CHAPTER 3

Scoping the Possibilities: Student Preferences towards Open Textbooks Adoption for E-Learning

Deepak Prasad, Tsuyoshi Usagawa

Department of Computer Science and Electrical Engineering, Graduate School of Science and Technology, Kumamoto University, Kumamoto, Japan

ABSTRACT

Many universities have begun implementing e-Learning systems due to their low cost. However, publishers of expensive textbooks stand in the way of e-Learning's ability to provide a cost-effec- tive educational delivery model. While many universities aim to overcome this opposition and replace traditional publishers' textbooks with free open textbooks, such plans cannot be executed successfully unless students are open to their use. As such, a study into students' preferences towards open textbook adoption is vital to provide clear indication as to their opinions regarding open textbook use. Thus, this study conducted a study of University of the South Pacific (USP) students' preferences towards open textbook adoption for e-Learning using a survey administered during Semester 2, 2013 which generated 1077 responses. Areas examined include: Impacts of textbook costs on students' academic careers; preferences towards open textbook adoption; perceived barriers to and motivations for adoption of open textbooks; and preferred digital features and reading devices. Results show that textbook prices adversely impact students. Furthermore, a high level of acceptance towards the adoption of open textbooks was found. The study revealed that the preference for reading printed material was the highest rated barrier to open textbook adoption, while the free availability of open textbooks was rated the greatest motivator. Study find- ings are being used to inform efforts to develop open textbooks at the USP and may assist other universities seeking to start similar projects.

KEYWORDS

Open Textbook, Digital Textbook, Student Preferences, Adoption, E-Learning, The University of the South Pacific

1. INTRODUCTION

E-learning is being adopted by many universities throughout the world as a cost-effective educational delivery model for expanding and widening access to higher education for all. The University of the South Pacific (USP), a dual-mode regional university co-owned by 12 Pacific island countries (Cook Islands, Fiji, Kiribati, Marshall Islands, Nauru, Niue, Samoa, Solomon Islands, Tokelau, Tonga, Tuvalu and Vanuatu) is one such university, particularly due to its unique mission: To deliver cost-effective quality education online to students living in the remote Pacific islands spread across 33 million square kilometers of ocean more than three times the size of Europe. In an effort to fulfill its mission and meet the critical needs of remote students, the USP is currently working towards developing 10 online learning programs by 2018 (University of the South Pacific, 2013: p. 13) .

Unfortunately, the constant rise in the price of publisher textbooks poses a major hurdle in the university's ambition to provide cost-effective tertiary education to students studying in the most remote places in the world. In fact, the cost of publishers' textbooks has risen a staggering 812% over the last 36 years (Perry, 2012), with growing evidence revealing that many students are unable to afford textbooks (Allen, 2011; Graydon, Urbach-Buholz, & Kohen, 2011; Rube, 2005). A recent survey (Senack, 2014) of 2039 university students reported that 65% of students decided not to purchase textbooks due to their expense. Most significantly, numerous studies have found that students who do not have their own textbook copy frequently fall behind, compromising their learning outcomes and increasing their probability of failing their course (Allen, 2011; Graydon et al., 2011; Morris-Babb & Henderson, 2012; Senack, 2014).

Though rising prices continue to affect students to the extent where some forgo purchasing textbooks, an interesting paradox exists: textbooks are often a prescribed component of courses offered at the USP. Prasad and Usagawa (2014) estimate that a USP student spends close to $200 on textbooks each semester, bringing annual textbook costs to $400 per student. Indeed, research shows that students are opting out of purchasing prescribed textbooks despite knowing that doing so would negatively affect their grades (Senack, 2014). Remarkably, prescribed textbooks account for approximately 75 to 90 percent of course discourse (Stein, Stuen, Carnine, & Long, 2001). With such a high percentage of textbook-based instruction and the fact that textbooks play a critical role in students' achievement, the need for a cost-effective solution to the problem of textbook affordability is urgently required at USP.

Recently, several textbook researchers (Allen, 2008b; Hilton & Wiley, 2011; Okamoto, 2013; Senack, 2014) have asserted the potential of open textbooks as a solution to expensive commercially published textbooks. Senack estimates that open textbooks could save students around $100 per course, a plausible assessment. For instance, British Columbia's open textbooks project has already saved students an average of $146 each on their textbook costs (Government of British Columbia, 2014). Most importantly, research has shown that students who study with open textbooks perform as well on tests as do their peers who use traditional textbooks (Wiley, Hilton, Ellington, & Hall, 2012). In the context of tertiary education, a traditional "textbook" is commonly understood as being instructional material used within tertiary education, delivered to the user on paper, in the form and binding produced and distributed by a publisher (Education for Change Ltd. & University of Stirling, 2003: p. 11). Open textbooks "are similar to traditional textbooks in terms of content; however, they are generally available for free in digital format, along with low-cost print copies" (Hilton, Gaudet, Clark, Robinson, & Wiley, 2013: p. 38). However, while open textbooks are digital textbooks, not all digital textbooks are open textbooks (i.e. free textbooks in digital formats). The term "digital" in "digital textbooks" means a textbook that is available in digital (or electronic) format such as HTML, EPUB, MOBI, OPF or PDF. Normally, digital textbooks are "consumed on a screen rather than on paper" (Nelson, 2008: p. 42).

It is important to point out that textbooks in digital formats are not merely digitized replicas of printed text- books. With recent developments in new and affordable educational technologies, textbooks in digital forms in-creasingly enable positive impacts on publishing, delivery, learning and teaching. As such, open textbooks not only possess the strong advantage of being free, but they also offer further advantages over traditional printed textbooks such as:

- More features-open textbooks may include interactive learning functions such as bookmarks, highlighting, annotations, text searching, quizzes, and hyperlinks; multiple digital media such as text, pictures, audio, video, animation, and interactive simulation; and options to synchronize offline and online learning data, which may be used to analyze students' reading patterns to enable subsequent improvement of the text and pedagogical methods.

- Better accessibility-open textbooks may be developed in a range of formats such as Web, EPub, PDF. This results in reduced physical size and weight, enabling increased portability and mobility, and provides options to print, read online and download for offline reading on various hardware devices such as a dedicated hand- held device, a personal digital assistance (PDA), a mobile phone, or a desktop or laptop computer. The digi- tal format reduces production and distribution time, and consequently facilities expeditious availability of texts that further support access.

- Greater flexibility-open textbooks may be updated quickly and provide access to the latest content. It "could be updated, say, to incorporate new knowledge. It could be improved as students and teachers develop better ways of expressing concepts or ordering learning objects. It could be localized or customized for a variety of learners, whether in different cultures or at different levels of education" (Matkin, 2009: p. 3).

Despite the abovementioned benefits, the adoption of digital textbooks has reportedly been much slower than anticipated (Allen, 2008a; Guthrie, 2012; Lee, 2010; Oliveira, 2012; Thomas, 2007). The reasons for this slow adoption vary, but 3 principal reasons emerge from literature.

To begin with, numerous studies have shown that students' attitudes and preferences toward printed books were more positive than toward digital textbooks (Armatus, Holt, & Rice, 2003; Armstrong, 2008; Buzzetto- More, Sweat-Guy, & Elobaid, 2007; Folb, Wessel, & Czechowski, 2011; Levine-Clark, 2006; Li, Poe, Potter, Quigley, & Wilson, 2011; McKnight & Dearnley, 2003; Spencer, 2006; Woody, Daniel, & Baker, 2010). Cuillier and Dewland (2014), for example, in their pilot study of digital textbook integration into a business course, found that almost 64% (23 out of 36) survey respondents preferred to read textbooks in print. Several themes consistently appear in literature on preferences for print format. These include less visual fatigue; more retention; greater comfort and relaxation; less anxiety; faster reading; and dependability and ease of use. However, unlike most commercial digital textbooks, open textbooks can be printed, thus resolving the concerns of those who prefer print format.

Next is the issue of digital textbook compatibility with hardware and software (Lee, Messom, & Yau, 2013). A digital textbook has three elements: the digital textbook content file, software to read the file, and a hardware device to view it on (Cavanaugh, 2002). Compatibility depends on these three elements; in other words, "you need the right software to read the right format, and you need that software installed on a compatible hardware device" (Nix, 2010: para. 9). Actually, digital textbook content in various file formats should be accessible on various hardware devices (for example, PC, laptop, PDA, or a dedicated hand-held reading device) to suit stu- dents' reading preferences. However, not all devices are compatible with all digital textbook format options (Buell, 2013: para. 2). Anuradha and Usha (2006) claim that digital textbook adoption rates have been slow because of its availability over disparate formats which are often "incompatible and non-interoperable" (p. 49). In the same vein, Landoni and Hanlon (2007) and Nelson (2008) acknowledged the possibility that compatibility problems may be a major force in slowing digital textbook adoption. This view has been confirmed by several studies. For example, in their interview study of 180 students and 20 academic staff members eliciting opinions about the challenges facing digital textbook adoption across schools in Bandar Sunway in Malaysia, Lee et al., (2013) found the most common perceptions about digital textbook use were related to difficulties associated with compatibility of digital books technology. Teachers in their study commented, "E-textbook reader device and content format incompatibility will be a problem," and "[t]he 'format' war for e-textbooks is a hurdle that must be overcome", while one student commented that "I won't buy anything if it's not compatible with all of my devices" (p. 35).

Third is the problem driven by students' lack of contentment with digital textbook features. Brahme and Gabriel (2012) surveyed graduate students' experiences and preferences regarding digital textbooks. They reported that lack of digital textbook features such as note-taking and highlighting caused frustration to 63% of their participants. Brahme and Gabriel (2012) asserted that students are often frustrated with digital textbook features that do not satisfy their needs. In another major study of how students use browser-based digital books (not necessarily textbooks), Berg, Hoffmann, and Dawson (2010) found that many students were frustrated with the structure and functionality of digital book features as they did not function according to their expectation. The authors concluded that, while interactive features are an advantage digital books have over printed books, these features must function well and be easily understood by the users for digital books to be more widely accepted.

To conclude, the three root causes for the current slow adoption of digital textbooks are: Greater preferences for reading printed textbooks over digital ones; incompatibility with students preferred reading device(s); and incongruence of digital textbook features with students' expectations. Taken together, these findings indicate that successful adoption of digital textbooks is primarily dependent on student preferences.

How to apply the findings from these studies to the adoption and acceptance of open textbooks among stu- dents, however, remains unclear, as most previous studies have only focused on adoption of commercial digital textbooks, not open textbooks. Such findings are thus not applicable, especially since open textbooks are free and printable; cost savings and printing options may encourage students to choose open textbooks over tradi- tional textbooks. Moreover, open textbook publishing currently lacks the established vigorous editorial mechan- isms found in traditional publishing models, eliciting uncertainties about accuracy and reliability of the content (Educause, 2011), which may cause students to lean towards traditional textbooks. As such, students' prefe- rences towards open textbook adoption-as opposed to commercial digital textbooks-remain unclear.

At the moment, USP is investigating possibilities to integrate open textbooks to its online courses in an effort to provide students with more affordable, interactive and flexible textbooks. In order to implement open text- books effectively, it is vital for USP to craft clear strategies for the adoption of open textbooks for e-Learning. In this vein, prior consultation with students themselves is required as their preferences towards open textbook adoption and factors influencing their choices will provide valuable information in predicting their acceptance of open textbooks. Awareness of student preferences is also crucial since their academic success is at stake. With these concerns in mind, this study addressed four factors:

1. What are the implications of textbook costs on students' academic careers?

2. What are students' preferences and motivations for, and barriers against, open textbook adoption?

3.   What are students' most desired digital features in open textbooks?

4.   What are students' most preferred devices for accessing open textbooks?

2. METHODOLOGY

The positivist paradigm was adopted to answer the above research questions. A quantitative Web-based survey questionnaire partly making use of questions from 2012 Florida Student Textbook Survey (Florida Virtual Campus, 2012) was constructed in five sections: 1) demographics; 2) impacts of textbook costs; 3) open text- book adoption; 4) desired digital features; and 5) desired reading devices. Three experts verified content validity of the questionnaire, with the questionnaire modified based on their responses and comments. A conditional question was included in Section 3: If the respondent answered "yes for some of my courses" or "yes for all my courses" to open textbook adoption questions, they were taken to motivator question, while a "no", "maybe" or "undecided" response took them to barrier questions. A variety of question formats was used: forced choice, multiple choice, multiple select, rating, skip-logic, and Likert scale. The final questionnaire included 18 ques- tions. To ensure reliability of the final version, test-retest reliability with a two-week interval was conducted on 7 students. The results obtained were subjected to Cronbach's alpha test, and the overall reliability of the final questionnaire reached (r = 0.84), which was acceptable to proceed with the survey. This study was conducted in accordance with all the requirements of ethical considerations. A self-selected sampling method was employed to generate a sample of students consisting of full-time and part-time students registered with the USP for aca- demic year 2013. The questionnaire was distributed online via Google Forms and was made available for one month from mid-November to mid-December in 2013. All students were e-mailed regarding the anonymous survey, and one reminder was issued. E-mails that were returned as "undeliverable" were removed from the sample size. The total number of students able to see the invitation to complete the survey was approximately 13,000. 1138 questionnaires were received, and after rejecting 61 partially filled-in questionnaires, 1077 remained available for analysis. Thus, the response rate was 8%. The data gathered via Google Forms were exported to MS-Excel for analysis based on the research questions. The results of the study are discussed in the next section.

3. RESULTS

To ensure accurate interpretation, the Hilton et al., (2013: p. 38) definition of an open textbook as cited in the introduction to this paper was provided to respondents in the invitation e-mail to participate as well as in the survey. The data from the survey's quantitative questions were analyzed and are presented below.

3.1. Demographics

Of 1077 respondents, 45% were male ($n = 489$) and 55% were female ($n = 588$). The majority of the respondents, 90% ($n = 970$), were undergraduate students, with only 10% ($n = 107$) being postgraduate students. Of the total respondents, 82% ($n = 888$) were full-time and 18% ($n = 189$) were part-time students. 70% of the respon- dents were less than 25 years old, 26% were between 26 and 40 years old, and the remaining 4% represented age groups older than 41. The students were categorized by their disciplines based on three different faculties in the USP, with the majority of students ($n = 429$) from Faculty of Business and Economics followed by Faculty of Science, Technology and Environment ($n = 372$), and Faculty of Arts, Law and Education ($n = 276$).

3.2. Frequency of Buying Prescribed Textbooks

Students were asked to give an indication of how often they bought prescribed textbooks. As Table 1 demon- strates, 484 of the 1077 respondents reported purchases very frequently or frequently, 331 bought occasionally, 183 rarely bought, and 79 reported that they never purchased prescribed textbooks.

3.3. Number of Prescribed Textbooks Bought in Semester 2, 2013

All the respondents ($n = 1077$) were asked how many prescribed textbooks they purchased for Semester 2, 2013. 74% ($n = 793$) reported buying at least one prescribed textbook, while 284 (26%) students indicated that they did not purchase any textbook for Semester 2, 2013. Table 2presents a breakdown of textbooks purchased in Semester 2, 2013. As illustrated, 793 students purchased a total of 1970 books. The number of textbooks bought by an individual ranged from 1 to 5 textbooks, with the mode falling in 2.

3.4. Textbook Expenditure for Semester 2, 2013

Those students ($n = 793$) who reported textbook purchase were asked to estimate the total Fijian dollar (1FJD = 0.55 USD) amount of their purchase for Semester 2, 2013. Table 3 demonstrates the respondents' replies, in terms of frequency and percentage of students by expenditure category. According to the results, 69% of the students reported spending over FJD200 on textbooks during Semester 2, 2013.The three most common expendi- ture categories were FJD000-100 (31%), followed by FJD201-300 (22%), and, in third place, FJD101-200 range chosen by 17% (n = 132) of respondents. The most striking result to emerge from the data is that 30% (n = 237) of the respondents spent more than FJD300 on textbooks.

Table 1. Prescribed textbook purchase frequency.]

How often do you buy prescribed textbooks?	Frequency
Very frequently	191
Frequently	293
Occasionally	331
Rarely	183
Never	79

Table 2. Number of prescribed textbooks purchased.

How many prescribed textbooks did you buy for Semester 2, 2013?	Frequency	Percentage
None	284	26
One	186	17
Two	227	21
Three	216	20
Four	138	13
Five	26	2

3.5. Scholarship for Textbook Expenses

Those students (n = 793) who had purchased textbooks were asked to clarify whether they were on scholarship during Semester 2, 2013. Remarkably, 650 (82%) of the 793 students who had bought textbooks reported that they were on scholarship during the second semester of 2013. Scholarship recipients (n = 650) were then asked to indicate the percentage of their textbook costs covered by the scholarship. 22% reported that their scholarship did not cover any of the textbook costs, 23% said that all their textbooks costs was covered by the scholarship, and 57% indicated that a portion of their textbooks costs was covered by the scholarship (Table 4).

Table 3. Semester 2, 2013 textbook expenses.

How much did you spend on your textbooks for Semester 2, 2013?	Frequency	Percentage
FJD000-100	247	31
FJD101-200	132	17
FJD201-300	177	22
FJD301-400	106	13
FJD401-500	116	15
More than FJD500	15	2

3.6. Methods of Textbook Purchase

Students (n = 793) who had purchased textbooks were asked to select their methods of textbook purchase. Tak- ing into account that an individual would buy more than one text (see Table 2), multiple-select was allowed. The results, as shown in Table 5, indicate that the most common method of purchase was buying new, printed textbooks from the campus bookshop (64%), which was followed by buying used printed textbooks (21%). The least popular methods of purchase were buying digital textbook with permanent access (8%) and buying digital versions with temporary access (7%). The single most striking observation to emerge from the data comparison was that 85% of the students purchased printed textbooks, while only 15% opted for digital versions.

3.7. Reasons for Not Buying a Textbook

Those students (n = 284) who reported not purchasing a textbook were asked to select cause(s) from a prepared list of reasons for not doing so. The list of reasons, frequency, and percentage is presented in Table 6. As illustrated, the four most commonly cited reasons for not buying a textbook were unaffordability (42%), no textbooks prescribed (15%), using a classmate's copy (14%), and photocopying required chapters from the textbook (10%).

Table 4. Percentage of textbook expense covered by scholarship.

What percentage of your textbook expense was covered by scholarship for Semester 2, 2013?	Frequency	Percentage
None	140	22
Less than 25%	120	18
26% to 50%	130	20
51% to 75%	67	10
76% to 99%	46	7
All textbook expense	147	23

Table 5. Textbook purchase method.

For Semester 2, 2013, how did you purchase your textbooks? Please select all that apply.	Frequency	Percentage
I purchased new print versions from the campus bookshop.	707	64
I purchased used print versions from former students.	231	21
I purchased digital textbooks-temporary ownership license.	75	7
I purchased digital textbooks-permanent access.	90	8

Table 6. Reasons for not buying a textbook.

Which of the following reason(s) explain why you did not buy a textbook in Semester 2, 2013? Please select all that apply.	Frequency	Percentage
Not prescribed for the course(s) I took.	56	15
Too expensive: not able to afford it.	160	42
Borrowed the textbook from my classmates.	52	14
Borrowed the textbook from the campus library reserve shelf.	32	8
Photocopied the whole textbook.	22	6
Photocopied the required chapters from the textbook.	38	10
The textbooks were sold out in campus bookshop.	17	5

3.8. Textbook Cost Consequences

The respondents (n = 1077) were asked to rate the effects of high textbook costs on their academic career (see Table 7). As shown, students indicated that the high cost of textbooks has caused them to, frequently, occasionally, or seldom: not purchase prescribed textbook (65%), submit assigned activities late (57%), earn poor grade (56%), take fewer courses (44%), fail a course (39%), not register for a course (31%), drop a course (28%), or withdraw from a course (26%).

3.9. Actions Taken to Reduce Textbook Cost

Table 8 shows the various actions respondents (n = 1077) took in order to reduce costs of textbooks. The vast majority of the students (72.2%) reported taking one or more measures to reduce the costs of their textbooks. Among the 10 actions to reduce textbook costs, the 3 most frequently reported were: buying used copies from former students (81.7%), using a reserve copy from the campus library (81.2%), and sharing books with classmates (81.1%).

3.10. Intention to Adopt Open Textbooks

Table 9 presents the distribution of student responses on their willingness to use open textbooks in the future given the choice of free access to digital versions and/or print at your own cost. Of the 1077 students who responded to the survey, 69% (n = 743) said they would use open textbooks for some or all of their courses, while only 3% (36/1077) reported that they had no intention of using open textbooks. The combined total of 28%: "maybe" (21%) and "undecided" (7%), indicated respondents' indecisiveness on whether they would use open textbooks in the future.

3.11. Motivations for Adopting Open Textbooks

Only those students (n = 743) who intended to use open textbooks for some or all of their courses were asked to rate motives that influenced their decision. These students were asked to rate 10 motivator items on a 5-point Likert-type scale ranging from 0 to 4for each of the motivator items, where 0 represented a motive that had "no influence" on their decision and 4 represented a "very influential" motive. Each score on the Likert scale was then converted to a mean motivator score (0 = 0, 1 = 25, 2 = 50, 3 = 75, 4 = 100), so that higher scores indicated stronger motivation. Table 10 illustrates the rank order and motivator strength (mean) of these 10 motivators. As illustrated, the motivator strength ranged from a high of 89.0 for the item 'They are freely available' to a low of 58.1 for the item "They are visually appealing".

3.12. Barriers to Adopting Open Textbooks

Those indecisive students (n = 298) and the students who were not interested (n = 36) in using open textbooks were asked to rate the strength of each barrier from a set of 6 potential barriers to open textbooks adoption on a Likert-type scale with scores ranging from 0 (no influence) to 4 (very influential). All the scores were trans- formed to a 0 - 100 scale (0 = 0, 1 = 25, 2 = 50, 3 = 75, 4 = 100). Table 11 shows the rank order and barrier strength (mean) of these items.

Table 7. Textbook cost consequences.

Consequence	Never		Seldom		Occasionally		Frequently	
	n	%	n	%	n	%	n	%
Take a fewer courses.	603	56	151	14	129	12	194	18
Not to register for a specific course.	743	69	129	12	97	9	108	10
Drop a course.	775	72	140	13	75	7	86	8
Withdraw from a course.	797	74	118	11	75	7	86	8
Earn a poor grade because I could not afford the textbook.	474	44	183	17	194	18	226	21
Fail a course because I could not afford the textbook.	657	61	162	15	118	11	140	13
Submit my assigned activities late because I did not have the prescribed textbook.	463	43	151	14	194	18	269	25
Not purchase the required textbook.	377	35	129	12	162	15	409	38

Table 8. Actions taken to reduce textbook costs.

Action	Yes		No	
	n	%	*n*	%
Make no attempt to reduce textbook cost.	299	27.8	778	72.2
Share books with classmates.	873	81.1	204	18.9
Buy used or new books online from a source other than the campus bookshop.	657	61.0	420	39.0
Buy a digital version of a textbook.	585	54.3	492	45.7
Buy used copies from former students.	880	81.7	197	18.3
Do not purchase the prescribed textbook.	790	73.4	287	26.6
Use a reserve copy from the campus library.	875	81.2	202	18.8
Photocopy only the chapters needed for the course.	798	74.1	279	25.9
Photocopy the whole textbook.	524	48.7	553	51.3
Sell used books.	753	69.9	324	30.1

3.13. Preferred Digital Features

To elucidate the most preferred digital features in open textbooks, all those students (n = 743) who had indicated their willingness to use open textbooks for some or all their courses were asked to rate the preference of 10 fea- tures on a 5-point Likert scale (where a score of "0" was "least preferred" and "5" "most preferred"). Mean was calculated by converting all the scores to a 0 - 100 scale, in a manner that higher mean scores indicate greater preference. Table 12 depicts the rank and mean of the 10 digital features provided in the list.

Table 9. Intention to use open textbooks.

Intention to use open textbook in the future	Frequency	Percentage
No	36	3
Maybe	227	21
Undecided	71	7
Yes, for some of my courses	345	32
Yes, for all my courses	398	37

Table 10. Motivators to use open textbooks.

Rank	Motivators	Motivator strength (mean)
1	They are freely available.	89.0
2	They can be self-printed and read.	87.2
3	They have interactive features that are not available in printed books (e.g. search functions).	83.5
4	They have features which are in printed books but easier to use in digital version (e.g. hyperlinked table of contents).	78.3
5	They have greater mobility and are convenient to carry.	77.6
6	They are easily portable and can be read on various reading devices.	76.6
7	They can be downloaded to a personal device and read offline.	75.7
8	They can be read online.	74.3
9	They are easy to use.	70.3
10	They are visually appealing.	58.1

Table 11. Barriers to use open textbooks.

Rank	Barriers	Barrier strength (mean)
1	I prefer reading printed materials.	71.3
2	I am afraid digital textbooks may not be compatible with all my reading devices.	63.7
3	I am worried about the quality of content.	63.3
4	I do not have experience in using digital textbooks.	62.7
5	I am not confident with using digital textbooks.	56.6
6	I do not have access to technology required to take advantage of digital textbooks.	43.6

3.14. Preferred Reading Devices

Table 13 shows preferences for reading devices for open textbooks as expressed by those students (n = 743) who were in favor of using open textbooks for some or all their courses in the future. As shown, from a pre-specified list of 4 common reading devices, the largest proportion of students (64%) preferred laptops; when combined with other mobile devices such as tablets (15%) and mobile phones (2%), 81% of the students pre- ferred to access open textbooks through a mobile device.

Table 12. Preferred digital features.

Rank	Features	Mean
1	Hyperlinked table of contents	81.5
2	Adding notes	81.2
3	Bookmarking	80.5
4	Searching within the textbook	80.3
5	Provides links to websites	80.1
6	Copying and pasting	79.0
7	Incorporates videos, figures, diagrams, or images	77.9
8	Highlighting	76.5
9	Printing	71.6
10	Text size control	67.7

Table 13. Most preferred reading device.

Device	Frequency	Percentage
Laptop	476	64
Desktop computer	141	19
Tablet (iPad, Galaxy, other)	111	15
Mobile phone	15	2

4. DISCUSSION

This study surveyed three factors affecting USP students' prescribed textbook buying behaviors and their preferences towards open textbooks adoption in place of traditional publisher texts as prescribed textbooks for their online courses: Cost, digital features, and preferred reading devices. The current study found that 814 of the 1077 students very frequently, frequently or occasionally bought prescribed textbooks. Similarly, in an earlier study by Carpenter, Bullock, and Potter (2006), almost three quarters of students reported buying textbooks that had been prescribed to them. As a result of their research, they concluded that the probability of a student buying a prescribed textbook depends of the strength of the endorsement given by their course lecturers. This also seems to a determinant of USP students' purchase decisions for prescribed textbooks. Of the 1077 students who took part in the survey, the majority, 74% (n = 793), indicated buying at least one prescribed textbook during Semester 2, 2013. The purchasing quantity ranged from 1 to 5 prescribed textbooks, with the mode falling in 2. In part, this wide array of purchases may be due to the number of courses taken by each student. The textbook expenditure reported for Semester 2,

2013, ranged from FJD000-100 category to over FJD500, with approximately one-third of those who made purchase spent over FJD300. Interestingly, 650 out of 793 students who bought textbooks were on scholarship. From the 650 scholarship recipients, 23% received full funding for textbook expenses, 57% got partial grant, and only 22% did not receive any financial aid for their textbook expenses. What is surprising is that amongst those students (n = 793) who purchased textbooks, the majority of them (64%) received some financial grant towards their textbook expenses. The data suggest that students who do not receive any financial support towards textbook expenses are less likely to purchase textbooks. It is perhaps not surprising that 42% of those students who reported not buying a textbook said they had not purchased one or more textbooks because the cost was too high. Three other common reasons for not buying a textbook included: textbook not required for a course (15%), borrowing a classmate's textbook (14%), and photocopying required chapters from the textbook (10%). The five most detrimental effects expensive textbooks had on students' academic careers were revealed as inability to purchase prescribed textbooks, late submission of assigned activities due to no personal copy of prescribed text, poor grades, fewer courses, and course failure. This clearly indicates that high textbook costs have a negative impact of students' academic careers. Given the adverse effects of expensive textbooks, students were asked if they did anything to reduce their textbook expenses. It is not surprising that a large majority (72.2%) of respondents indicated that they used more than one strategy to save money on textbook cost. The most popular method reported was buying used textbooks, with 81.7% of respondents indicating that they utilized this method. This is in line with the recent finding by the AdHoc Senate Committee on Student Textbook Savings, where 81.4% of the respondents reported buying used textbooks (AdHoc, 2014). The three most popular methods reported were sharing classmate's books, using a library copy, and buying used textbooks.

In response to the question of whether the students are willing to use open textbooks in the future for their online courses, the majority responded in the positive, "Yes, for some of my courses" (32%) or "Yes, for all my courses" (37%). These results showed that free availability of open textbooks was the highest rated motive behind these willing students' intention to use open textbooks in the future. This was expected since reduced cost is the most commonly reported benefit of open textbooks (Hilton, Robinson, Wiley, & Ackerman, 2014; Wiley et al., 2012). The next most highly rated motivator was ability to self-print. Again, this comes as no surprise as most previous studies have consistently found that students prefer to read printed rather than on-screen materials (Buzzetto-More et al., 2007; Spencer, 2006; Woody et al., 2010). Interactive features, which are not available in printed books, were ranked at third position. Concerning barriers to open textbooks adoption: Preference for reading printed materials, concerns about compatibility, and worries about quality of content were ranked as the top-three barriers to adoption of open textbooks by those students who were not in favor or undecided about using open textbooks. The top two barriers are commonly cited barriers to adoption of textbooks in digital formats (as identified in the introduction to this paper); however, concerns regarding quality of content have also emerged as a strong barrier that may overshadow open textbooks free of price. In connection with quality, Cragun (2007) asked his students if they preferred a more developed textbook that costs money or a free textbook that covered just what they needed to know for the course. Students preferred using a free textbook. Cragun remarked that free does not always mean the open textbook is good, but he recognized quality is important and that more people collaborating in development will help ensure this. In the final part of the paper, he writes: "My students clearly liked the text, despite its flaws, but this was likely due-in large part-to the text being free" (Cragun, 2007: p. 11). The results of Cragun's study showed that students liked the zero dollar cost and did not worry that the textbook was incomplete as long the textbook covered the course learning contents.

With regard to student preferences for digital features, hyperlinked table of contents was the most preferred feature with a mean of 81.5. The next most preferred features, in descending order, were adding notes, bookmarking, searching within the textbook, and links to websites (all these features has mean above 80). Several research studies (Behler & Lush, 2010; Chong, Lim, & Ling, 2009; Cuillier & Dewland, 2014) have shown that these features were atop the wish list of students. That the majority of students in this and other studies strongly prefer these features substantiates making these features richer and standardized across digital textbook technology. According toBehler and Lush (2010), digital textbook features are far from where they need to be to allow digital textbooks to replace traditional books. Concurring with Behler and Lush (2010), Philip and Moon (2013) purported that there is a need for significant improvements in the features of digital textbooks for them to be widely adopted. This further validates the call for the development of better and consistent digital textbook features.

On the question of students' most preferred device for reading open textbooks, most (64%) indicated laptops, 19% desktop computers, 15% tablets, and 2% mobile phones; consequently, 81% of the students preferred to access open textbooks through a mobile device. This finding is consistent with recent studies of digital textbooks such as Rockinson-Szapkiw, Courduff, Carter, and Bennett (2013), Cuillier and Dewland (2014), and Hwang, Kim, Lee, and Kim (2014). The current study found that the least preferred device was mobile phone. This finding is not surprising given results from prior studies that have found this type of technology has not shown great popularity for the purpose of digital textbook reading (Croft & Davis, 2010; Zimerman, 2011). Some authors (e.g. Zimerman, 2011) have speculated that, with appropriate software, mobiles phones could double as a viable reading device, particularly since most students have mobile phones. As such, it is important to develop platform-independent open textbooks that are accessible through any reading device with a simple browser-based interface. Finally, the response rate to the survey was limited and much lower than anticipated, which might limit the generalizability of findings to the target population. However, the study was feasible to conduct as a starting point for further work.

5. CONCLUSION AND FUTURE DIRECTIONS

This study was concerned with USP students' preferences towards adoption of open textbooks for online courses; however, the results should be applicable also to other USP course modes (blended, print and face-to-face). The results of the current study are extremely encouraging, with a good percentage (69%) of the surveyed students indicating their willingness to use open textbooks for some or all of their courses, though not all students were keen in using open textbooks. But like any other technological innovation in education, with continued usage, students are likely to gradually get more familiar and confident in using textbooks in digital formats. Hoseth and Merinda (2012) in their paper similarly concluded that participants in their study clearly expressed the need to adapt to change and switch to digital textbook formats, and successively becoming more familiar with them over time. This is in consonance with Chou (2014: p. 16) when he concludes that, "When students are given time and opportunities to read e-books, they are likely to develop e-book reading habits. It would be a pity if we gave up providing students e-books merely because of their initial negative attitudes."

In consideration of the above findings, the study recommends the viability of open textbooks. Expensive traditional textbooks will need to be replaced with open textbooks as prescribed textbooks for USP courses. Clearly, this will be no mean feat. Future work will involve: 1) Identifying USP teachers willing to collaborate with the primary author in the development of custom-built OER derived open textbooks; 2) Developing an open textbook learning analytics system; 3) Piloting an open textbook over a semester; 4) Evaluating an open textbook and the analytics system; 5) Documenting the development processes; and 6) Disseminating the outcomes through conferences and journals as well as focused events and workshops within USP. Considering the convincing results and suggested research directions, there is hope that open textbooks will replace their traditional counterparts as prescribed texts and in process benefit USP students.

ACKNOWLEDGEMENTS

Part of this work was supported by Grant-in-Aid for Scientific Research 25280124. Deepak Prasad is a member of the Global OER Graduate Network (GO-GN), http://www.ou.nl/go-gn.

REFERENCES

1. AdHoc Senate Committee on Student Textbook Savings (2014). Recommendations to Save on Student Textbooks Costs.http://www.lib.utah.edu/pdf/TextbookSavingsReport_FINAL.pdf

2. Allen, N. (2008a). Course Correction: How Digital Textbooks Are off Track and How to Set Them Straight (pp. 1-21).http://www.immagic.com/eLibrary/ARCHIVES/GENERAL/ST_PIRGS/S080826A.pdf

3. Allen, N. (2008b). Digital Textbooks: A Student Perspective. New England Journal of Higher Education, 23, 32.

4. Allen, N. (2011). High Prices Prevent College Students from Buying Assigned Textbooks. Student PIRGs. http://www.studentpirgs.org/news/ap/high-prices-prevent-college-students-buying-assigned-textbooks

5. Anuradha, K. T., & Usha, H. S. (2006). Use of E-Books in an Academic and Research Environment: A Case Study from the Indian Institute of Science. Program, 40, 48-62.http://dx.doi.org/ 10.1108/00330330610646807

6. Armatus, C., Holt, D., & Rice, M. (2003). Impacts of an On-Line-Supported, Resource-Based Learning Environment: Does One Size Fit All? Distance Education, 24, 140-158.

7. Armstrong, C. (2008). Books in a Virtual World: The Evolution of the E-Book and Its Lexicon. Journal of Librarianship and Information Science, 40, 193-206.http://dx.doi.org/10.1177/0961000608092554

8. Behler, A., & Lush, B. (2010). Are You Ready for E-Readers? The Reference Librarian, 52, 75-87. http://dx.doi.org/10. 1080/02763877.2011.523261

9. Berg, S. A., Hoffmann, K., & Dawson, D. (2010). Not on the Same Page: Undergraduates' Information Retrieval in Electronic and Print Books. The Journal of Academic Librarianship, 36, 518-525. http://dx.doi.org/10.1016/j.acalib.2010.08.008

10. Brahme, M., & Gabriel, L. (2012). Are Students Keeping up with the E-Book Evolution? Are E-Books Keeping up with Students' Evolving Needs? Distance Students and E-Book Usage, a Survey. Journal of Library & Information Services in Distance Learning, 6, 180-198. http://dx.doi.org/10.1080/1533290X.2012.705109

11. Buell, C. (2013). E-Textbook Matchmaking: Assessing Compatibility between Hardware and Software. http:// edcetera.rafter. com/e-textbook-matchmaking-assessing-compatibility-between-hardware-and-software/

12. Buzzetto-More, N., Sweat-Guy, R., & Elobaid, M. (2007). Reading in a Digital Age: E-Books Are Students Ready for This Learning Object? Interdisciplinary Journal of Knowledge and Learning Objects, 3, 239-250.

13. Carpenter, P., Bullock, A., & Potter, J. (2006). Textbooks in Teaching and Learning. Brookes eJournal of Learning and Teaching, 2, No. 1.http://bejlt.brookes.ac.uk/paper/textbooks_in_teaching_and_learning-2/

14. Cavanaugh, T. (2002). Ebooks and Accommodations: Is There the Future. Teaching Exceptional Children, 34, 56-61.

15. Chong, P. F., Lim, Y. P., & Ling, S. W. (2009). On the Design Preferences for Ebooks. IETE Technical Review, 26, 213-222. http://dx.doi.org/10.4103/0256-4602.50706

16. Chou, I. C. (2014). Reading for the Purpose of Responding to Literature: EFL Students' Perceptions of E-Books. Computer Assisted Language Learning, 1-20.http://dx.doi.org/10.1080/09588221.2014.881388

17. Cragun, R. T. (2007). The Future of Textbooks? Electronic Journal of Sociology.http://www.sociology.org/content/2007/_cragun_futureoftext books.pdf

18. Croft, R., & Davis, C. (2010). E-Books Revisited: Surveying Student E-Book Usage in a Distributed Learning Academic Library 6 Years Later.

Journal   of   Library   Administration,   50,   543-
569. http://dx.doi.org/10.1080/01930826.2010.488600

19. Cuillier, C. A., & Dewland, J. C. (2014). Understanding the Key Factors
    for E-Textbook Integration into a Business Course: A Case Study.
    Journal   of   Business   &   Finance   Librarianship,   19,   32-
    60. http://dx.doi.org/10.1080/08963568.2013.824338

20. Education for Change Ltd., & University of Stirling (2003). A Strategy
    and Vision for the Future for Electronic Textbooks in UK Further and
    Higher   Education.http://www.jisc.ac.uk/   uploaded_documents/
    Annex_E_E_Textbooks_Strategy_final_report.pdf

21. Educause (2011). 7 Things You Should Know about Open Textbook
    Publishing.http://net.educause.edu/ir/library/pdf/eli7070.pdf

22. Florida Virtual Campus (2012). 2012 Florida Student Textbook Survey.
    Tallahassee.http://www.openaccesstextbooks.org/pdf/2012_Florida_Stu
    dent_Textbook_Survey.pdf

23. Folb, B. L., Wessel, C. B., & Czechowski, L. J. (2011). Clinical and
    Academic Use of Electronic and Print Books: The Health Sciences
    Library System E-Book Study at the University of Pittsburg. Journal of
    the   Medical   Library   Association: JMLA,   99,   218-
    228.http://dx.doi.org/10.3163/1536-5050.99.3.009

24. Government of British Columbia (2014). Students Saving Money with
    Open   Textbooks.   http://www.newsroom.gov.bc.ca/2014/01/students-
    saving-money-with-open-textbooks.html

25. Graydon, B., Urbach-Buholz, B., & Kohen, C. (2011). A Study of Four
    Textbook Distribution Models. Educause Quarterly, 34, 1-11.http://
    www.educause.edu/ero/article/study-four-textbook-distribution-models

26. Guthrie, K. M. (2012). Will Book Be Different? Journal of Library
    Administration,   52,   353-369. http://dx.doi.org/10.1080/   01930826.
    2012.700805

27. Hilton, J. L., & Wiley, D. (2011). Open-Access Textbooks and Financial
    Sustainability: A Case Study on Flat World Knowledge. The

International Review of Research in Open and Distance Learning, 12, 18-26.

28. Hilton, J. L., Gaudet, D., Clark, P., Robinson, J., & Wiley, D. (2013). The Adoption of Open Educational Resources by One Community College Math Department. The International Review of Research in Open and Distance Learning, 14, 37-50.

29. Hilton, J. L., Robinson, T. J., Wiley, D., & Ackerman, J. D. (2014). Cost-Savings Achieved in Two Semesters through the Adoption of Open Educational Resources. International Review of Research in Open Distance Learning, 15, 67-84.

30. Hoseth, A., & Merinda, M. (2012). Perspectives on E-Book from Instructors and Students in the Social Sciences. Reference & User Services Quarterly, 51, 278-288.http://dx.doi.org/10.5860/rusq.51n3.278

31. Hwang, J. Y., Kim, J., Lee, B., & Kim, J. H. (2014). Usage Patterns and Perception toward E-Books: Experiences from Academic Libraries in South Korea. The Electronic Library, 32, 522-541. http://dx.doi.org/10.1108/EL-11-2012-0150

32. Landoni, M., & Hanlon, G. (2007). E-Books Reading Groups: Interacting with E-Books in Public Libraries. The Electronic Library, 25, 599-612.http://dx.doi.org/10.1108/02640470710829578

33. Lee, H. J., Messom, C., & Yau, K. A. (2013). Can an Electronic Textbooks Be Part of K-12 Education? Challenges, Technological Solutions and Open Issues. The Turkish Online Journal of Educational Technology, 12, 32-44.

34. Lee, M. C. (2010). Explaining and Predicting Users' Continuance Intention toward E-Learning: An Extension of the Expectation-Confirmation Model. Computers & Education, 54, 506-516. http://dx.doi.org/10.1016/j.compedu.2009.09.002

35. Levine-Clark, M. (2006). Electronic Book Usage: A Survey at the University of Denver. Collection Building, 26, 7-14. http://dx.doi.org/10.1108/01604950710721548

36. Li, C., Poe, F., Potter, M., Quigley, B., & Wilson, J. (2011). UC Libraries Academic E-Book Usage Survey, Springer E- Book Pilot Project.http://www.cdlib.org/services/uxdesign/docs/2011/academic_ebook_usage_survey.pdf

37. Matkin, G. W. (2009). Open Learning: What Do Open Textbooks Tell Us about the Revolution in Education? In Center for Studies in Higher Education, Research & Occasional Paper Series: CSHE.1.09 (pp. 1-8). Berkeley, CA: University of California.http://cshe.berkeley. edu/sites/default/files/shared/publications/docs/ROPs-Matkin-OpenLearning-03-31-09.pdf

38. McKnight, C., & Dearnley, J. (2003). Electronic Book Use in a Public Library. Journal of Librarianship and Information Science, 35, 235-242.http://dx.doi.org/10.1177/0961000603035004003

39. Morris-Babb, M., & Henderson, S. (2012). An Experiment in Open-Access Textbook Publishing: Changing the World One Textbook at a Time. Journal of Scholarly Publishing, 43, 148-155. http://dx.doi.org/10.3138/jsp.43.2.148

40. Nelson, M. R. (2008). E-Books in Higher Education: Nearing the End of an Era of Hype? Educause Review, 43, 40-56.

41. Nix, L. (2010). Ebook Formats: The Basics.http://goldenorbcreative. wordpress.com/2010/09/02/ebook-formats-the-basics/

42. Okamoto, K. (2013). Making Higher Education More Affordable, One Course Reading at a Time: Academic Libraries as Key Advocates for Open Access Textbooks and Educational Resources. Public Services Quarterly, 9, 267-283.http://dx.doi.org/10.1080/15228959.2013.842397

43. Oliveira, S. M. (2012). E-Textbooks Usage by Students at Andrews University: A Study of Attitudes, Perceptions, and Behaviors. Library Management, 33, 536-560.http://dx.doi.org/ 10.1108/01435121211279894

44. Perry, M. (2012). The College Textbook Bubble and How the "Open Educational Resources" Movement Is Going up against the Textbook

Cartel. American Enterprise Institute.http://www.aei-ideas.org/2012/12/the-college-textbook-bubble-and-how-the-open-educational-resources-movement-is-going-up-against-the-textbook-cartel/

45. Philip, G. C., & Moon, S. Y. (2013). An Investigation of Student Expectation, Perceived Performance and Satisfaction of E-Textbooks. Journal of Information Technology Education: Innovations in Practice, 12, 287-298.

46. Prasad, D., & Usagawa, T. (2014). Towards Development of OER Derived Custom-Built Open Textbooks: A Baseline Survey of University Teachers at the University of the South Pacific. The International Review of Research in Open and Distance Learning, 14, 226-247.

47. Rockinson-Szapkiw, A. J., Courduff, J., Carter, K., & Bennett, D. (2013). Electronic versus Traditional Print Textbooks: A Comparison Study on the Influence of University Students' Learning. Computers & Education, 63, 259-266.http://dx.doi.org/10.1016/j.compedu.2012.11.022

48. Rube, K. (2005). Ripoff 101: How the Publishing Industry's Practices Needlessly Drive up Textbook Costs: A National Survey of Textbook Prices (2nd ed.). Washington DC.http://www.studentpirgs.org/ sites/student/files/reports/ripoff-101-2nd.pdf

49. Senack, E. (2014). Fixing the Broken Textbook Market: How Students Respond to High Textbook Costs and Demand Alternatives. Washington DC.http://www.washpirg.org/sites/pirg/files/reports/1.27.14 Fixing Broken Textbooks Report.pdf

50. Spencer, C. (2006). Research on Learners' Preference for Reading from a Printed Text or from a Computer Screen. Journal of Distance Education, 21, 33-50.

51. Stein, M., Stuen, C., Carnine, D., & Long, R. (2001). Textbook Evaluation and Adoption. Reading & Writing Quarterly, 17, 5-23. http://dx.doi.org/10.1080/105735601455710

52. Thomas, S. E. (2007). Another Side of the E-Book Puzzle. Indiana Libraries, 26, 39-45.

53. University of the South Pacific (2013). Strategic Plan 2013-2018. Suva.http://www.usp.ac.fj/index.php?id=12556

54. Wiley, D., Hilton, J. L., Ellington, S., & Hall, T. (2012). A Preliminary Examination of the Cost Savings and Learning Impacts of Using Open Textbooks in Middle and High School Science Classes. The International Review of Research in Open and Distance Learning, 13, 262-276.

55. Woody, W. D., Daniel, D. B., & Baker, C. A. (2010). E-Books or Textbooks: Students Prefer Textbooks. Computers & Education, 55, 945-948.http://dx.doi.org/10.1016/j.compedu.2010.04.005

56. Zimerman, M. (2011). E-Readers in an Academic Library Setting. Library Hi Tech, 29, 91-108. http://dx.doi.org/10.1108/07378831111116930

CHAPTER 4

From Workshop to E-Learning: Using Technology-Enhanced "Intermediate Concept Measures" As a Framework for Pharmacy Ethics Education and Assessment

Cicely Roche [1,*], Steve Thoma [2] and Joy Wingfield [3]

[1] School of Pharmacy; Panoz Building, Trinity College, College Green, Dublin 2, Ireland

[2] Educational Psychology; University of Alabama, Tuscaloosa, AL 35487, USA

[3] School of Pharmacy; Nottingham University, University Park Nottingham, Nottingham NG7 2RD, UK

ABSTRACT

Workshop analysis of scenarios or vignettes has traditionally been used to develop and demonstrate the moral reasoning underpinning professional decisions. However, in order to facilitate sufficiently individualized interaction to accommodate the assessment of student competencies related to decision-making through scenarios, such workshops are traditionally used with small groups. There are associated resource implications for the scheduling of sessions and implications for tutor time where large cohorts of students are targeted. In addition, the requirement that students be face-to-face is problematic when students are in practice placements that are geographically removed. This paper demonstrates how technology and an assessment tool, known as an "intermediate concept measure" (ICM), might help address these limitations. It introduces the background to ICMs and presents the ICM as a tool that has potential to support professional education. It also shares learning experienced by one pharmacist using ICMs in pharmacy education, provides an example of how a profession-specific ICM might be formatted, suggests how the methodology might be used in undergraduate and postgraduate education and provides samples of measurables that may be incorporated into evaluation and assessment systems; both for educational interventions delivered face-to-face or partly or entirely online. The limitations of the methodologies and suggestions for further research are included.

KEYWORDS

pharmacy ethics education; technology enhanced learning; moral reasoning competency development; intermediate concepts

1. INTRODUCTION

Workshop analysis of scenarios or vignettes has traditionally been used to develop and demonstrate the moral reasoning underpinning professional decisions, but, in order to facilitate sufficiently individualized interaction to accommodate the assessment of student competencies related to decision-making through scenarios, such workshops are traditionally used with small groups. There are associated resource implications for the scheduling of sessions and implications for tutor time where large cohorts of students are targeted. In addition, the requirement that students be face-to-face is problematic when students are in practice placements that are geographically removed. This paper shows how technology and an assessment tool, known as an "intermediate concept measure" (ICM), can help address these limitations. Intermediate concepts represent professional concerns, such as the professional "duty of care" and the patient's right to consent, confidentiality and "patient best interests", which are described in terms of guiding ethical standards for the professional [1,2]. They are referred to as "intermediate" in the context that they lie between the "surface level" rules, norms and codes governing the practice of the profession and the deeper level or "bedrock schema" reasoning processes, the development of which represents highly abstract moral judgment strategies. These "bedrock schema" serve as a default system that is activated when more automatic and context-specific interpretive systems fail or provide incomplete or inconsistent information.

ICMs were originally developed for dental students in order to assess moral judgments within a professional context, *i.e.*, the ability to both identify appropriate applications of the intermediate concepts in a profession-specific dilemma scenario and interpret how an individual's actions may affect the outcome of a dilemma caused by a conflict of concepts [2,3]. The design of professional ethics courses is often organized "around intermediate level concepts" [2] (p. 347). ICMs incorporate a short profession-specific "dilemma" scenario, written to incorporate ethical conflict(s), and are generally validated by a group of practitioners considered "experts" in the profession. The dilemma and series of action and justification options are presented in sequence (Appendix 1). The action and justification options proposed, which students must both rate and rank, include those with a focus on self-interest, maintaining rules and norms and acting in the patient's best interests or in societal interests [4,5]. The methodology includes the opportunity to have small groups that seek to agree on the preferred action and justification options from the proposed list.

Professional ethics education as proposed in "Moral Development in the Professions: Psychology and Applied Ethics" [6] incorporates moral sensitivity, moral reasoning, or "judgment", motivation, or "justification", and implementation, or "character", to act as intended, as interactive elements in the development of a professional, in what is known as the Four Component Model (FCM) of professional development [2,3,6]. These components are represented as interactive elements in the development of a professional, as depicted in Figure 1.

The FCM proposes that moral reasoning processing takes place at three levels:

i.   Developmental bedrock schemas, reflecting preferred decision-making schemas at an abstract level, as measured by a psychometric measure known as the Defining Issues Test (DIT) [4,7]. While the impact of educational interventions may be measured using a pre-post intervention design using the DIT as the measure [5], the discussion surrounding the measurement of the impact of professional ethics programs is beyond the scope of this paper.

ii.   Intermediate-level moral concepts are designed to cover a broad range of situations that require significant professional interpretation by participants in an educational intervention. Reasoning about intermediate concepts is, in part, a reflection of the individual's preferred approach to decision-making through dilemmas. The methodology outlined in this article draws from intermediate concepts and the FCM [2], but presents a format of ICM that is used to enhance the development of moral reasoning (Figure 1) in a manner that also accommodates some demonstration and assessment of related competencies.

iii. The more concrete, or surface level, processing incorporates rules (or legislation governing the practice of pharmacy) and codes of conduct or ethics, as generally included in professional ethics programs. However, the most difficult aspect of using professional codes as a framework for decision-making is that it is difficult to recognize when the endless variables in real-life scenarios, as included in a given dilemma scenario, are actually covered by the code. Practitioners typically engage surface-level moral reasoning when it accommodates the dilemma proposed and move to intermediate-level approaches only when a satisfactory action plan is not evident from the legislation or the professional Code of Conduct (CoC).

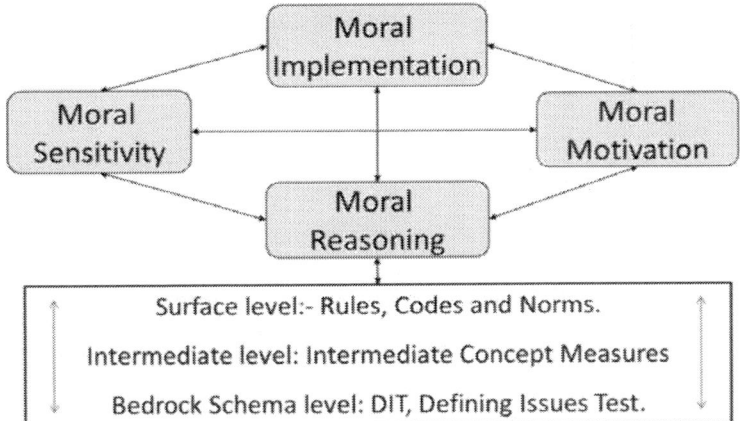

Figure 1. The Four Component Model of professional education [8]. DIT, Defining Issues Test.

The reality is that pharmacists regularly face ethical dilemmas, where there is a conflict of moral values creating a situation in which there is no obvious right or wrong answer [9], but where there are two or more options that are individually convincing, mutually exclusive and jointly demanding and none of which is necessarily in line with the letter of the law or a literal interpretation of the CoC. Recent research into the challenges encountered by pharmacists when faced with ethical dilemmas suggests that both undergraduates and practicing pharmacists would benefit from a structured means by which to build and maintain competencies related to dilemma review [8,9,10,11,12,13,14,15,16]. Latif's research [17,18], while it reports on pharmacists in the USA, rather than in Ireland, nonetheless indicates that community pharmacists are a rare exception to the expectation that moral reasoning competencies generally increase with age. The ability to reason through dilemmas to choose between available options and to justify decisions in a coherent manner should they be subjected to external scrutiny should be targeted as distinct competencies in pharmacy education [19].

Ethical development is a particularly relevant priority for contemporary pharmacy education, as research by Wingfield and colleagues has shown that "there is little research literature specifically addressing ethics in pharmacy practice and almost none addressing fundamental philosophical issues or values for pharmacy ethics" [20] (p. 2382). Their outcomes prioritize the teaching and assessment of "ethical competence" before practice and the development and updating of this competence in practicing pharmacists [20]. Research has presented many examples of potential conflict of interest dilemmas regularly faced by pharmacists (e.g., [8,12,13,14,15,16,18,21,22,23]), and scenarios presented in pharmacy-specific ICMs seek to capture examples of these dilemmas for teaching and learning purposes.

Research in other professions, e.g., dentistry [3], physical therapy [24] and business [25], indicates that even relatively short profession-specific educational programs can lead to significant improvements in moral development, especially when the design, development and delivery of the intervention is context appropriate. It is envisaged that ICM inclusion in pharmacy education would be constructively aligned with other aspects of professional programs [26,27], that rubrics that articulate grading expectations would be provided at the outset to guide both learning and assessment (samples are inAppendix 2 and Appendix 3) and that students would have been introduced to decision-making frameworks, such as Principlism [28] and value-based ethics [29,30], prior to engagement with the ICM, e.g., first-year students would have been introduced to both frameworks, through lectures and workshop activities, prior to engaging with ICMs. There is no suggestion that ICMs, presented in any of the configurations described in this article, could independently comprise a program of professional development.

The teaching and learning methodologies reported in this article use profession-specific ICMs to support technology-enhanced learning (TEL), entirely online or in a mix of face-to-face and online, or "blended learning", approaches, at undergraduate and post-graduate levels in pharmacy education. The general format by which ICMs are incorporated into educational interventions is as follows:

- Upon review of a previously unseen scenario (Appendix 1: Box 1), a student is forced to declare a "position" [31] when recording online her/his reflection or "independent review" of the dilemma therein (Appendix 1: Part 1). Constructivism, which proposes that learning is an active process, wherein new information is added to "prior knowledge", which may have been derived from personal experience, as well as formal teaching and learning [26,27,32,33], is the key learning theory employed.

- The rating and ranking of action (Appendix 1: Part 2a) and justification options (Appendix 1: Part 2b) challenges the student to revisit the scenario, where the options posed prompt consideration of a broader range of potential professional concepts and dilemmas, while also forcing a choice between less than ideal circumstances. The discovery of "differences between expert and novice groups enables the educator to judge individual performance against a valid standard" [2] (p. 358), and the validation process grounds the ICM in the thinking and reasoning of respected practitioner and educator members of the profession, making it more likely to engage those undertaking professional ethics programs.

- Part 3 of the process randomly allocates students to groups of five to seven members, who, having committed to individual choices regarding the rating and ranking of action and justification options offered, must agree and make a group decision regarding the ranking process within a defined time frame. This inevitably involves constructivism in the form of negotiation and active discussion, debate and persuasion, as the group seeks to complete the task by the deadline imposed. Peer debate forces deeper reflection on the decisions made.

As all contributions are recorded online, they collectively provide the student with a record of how she/he and the rest of the group reasoned through the dilemmas, provide an unambiguous declaration of group member's different individual "starting positions" and facilitate assessment of the demonstrated competencies by subsequent review of records in the virtual learning environment (VLE) [34].

To the authors' knowledge, pharmacy profession-specific ICMs have not been previously developed, and the presentation of ICMs in the online environment has not been reported in the literature. This article seeks to share the methodology that has been developed for the use of ICMs in pharmacy ethics education, to outline, using a specific example, how it is currently being piloted in various contexts in pharmacy education and to present some suggested outcome measures as available through the interrogation of the reporting functionality available in the VLE.

2. TECHNOLOGY-ENHANCED LEARNING: BLENDED LEARNING IN PHARMACY ETHICS EDUCATION

Technology has enhanced the potential of ICMs to support the use of the workshop analysis of scenarios or vignettes to demonstrate the moral reasoning behind professional decisions, while also helping to address staffing and resourcing issues where student numbers are large. This potential for enhancement may be considered to occur in each of the three parts of the ICM process, variously motivating student engagement in the process of dilemma review and discussion and providing the opportunity for the demonstration of competencies related, in particular, to team work. This process has been adapted to different formats to suit undergraduate and post-graduate programs. Examples of regularly used "adaptations" are outlined below:

- Undergraduate level: a blended learning approach aligned with a series of workshops, each student generally having online access during the workshop(s).

- ICM Part 1.

   Students review a previously unseen dilemma scenario (Appendix 1) and are required to answer the first question posed: "what is/are the ethical concepts in this scenario?" While this could be presented as a paper-based exercise, in which case, the workshop leader could either review answers and provide feedback at a later stage or invite samples of suggestions from students in order to influence subsequent discussion, access to a VLE by students engaged in face-to-face learning provides an opportunity for more comprehensive and interactive feedback to the group in the classroom or workshop setting. Where large numbers of students post contributions in real time, the use of word clouds as facilitated by Feinberg's "Wordle" software [35] where the size of the word is indicative of the number of times it appears in a document, can support timely feedback and formative assessment in a manner not possible for a single workshop leader with a paper exercise.

- The word cloud in Figure 2 was produced by collating all answers to Question 1 (Appendix 1) provided by one class group. This collation took less than one minute.

Figure 2. What is/are the ethical concepts in this scenario? [35]

- The word cloud (Figure 2) was then used as a focus through which to stimulate further discussion on the dilemma scenario. It has been the experience of the authors that students respond actively to data and feedback that represents their own (individual or peer-group) opinions, especially when it is provided while the memory of their own independent opinions is fresh, and this approach appears to support active engagement in the dilemma discussions that follow.

- Grading of Part 1 (Appendix 1) may, of course, be guided by the related rubric or guide (Appendix 2) and the use of the rubric online means that it is visible to the student as a guide when answering whatever questions are posed (the Sample 5 question format in Appendix 1). In context, a rubric is considered to be an assessment instrument that gives students information on how the tutor will be assessing their performance. Rubrics can increase transparency in assessment, because they make public the criteria for the judgment of student performance. The rubric may also guide grading and online feedback as a tutor or assessment design deems appropriate.

- ICM Part 2: Rate and rank the action and justification options (Appendix 1).

- Where a scenario and its lists of action and justification options are deemed to represent what an "expert" would suggest, student completion of the rating and ranking exercise under "live exam" conditions supports the claim that students are provided with the opportunity to demonstrate whether they can identify and define professional dilemmas in a given scenario when prompted to do so. In context, ICMs are considered representative of "expert" opinion, where validation has involved review by a group of pharmacists considered appropriately experienced. If a student ranks as most preferred an action option considered by experts to be the "worst" option, this suggests that the student has not recognized the dilemma. While it rarely happens, with either first- or fourth-year cohorts of pharmacy students, it prompts an individual review with the student(s) after the workshop or session, the identity of the individual involved being easily tracked through the VLE. In this manner the process supports the professionalization process in use in the degree program.

- Further potential to enhance learning may include the manner in which the students can be given immediate feedback as to how they collectively rated the options presented or ranked the three most and least preferred options. Functionality on the VLE permits visual presentation of the variety of opinion amongst the peer group, examples of which are presented for the first two action options (Appendix 1) in Figure 3.

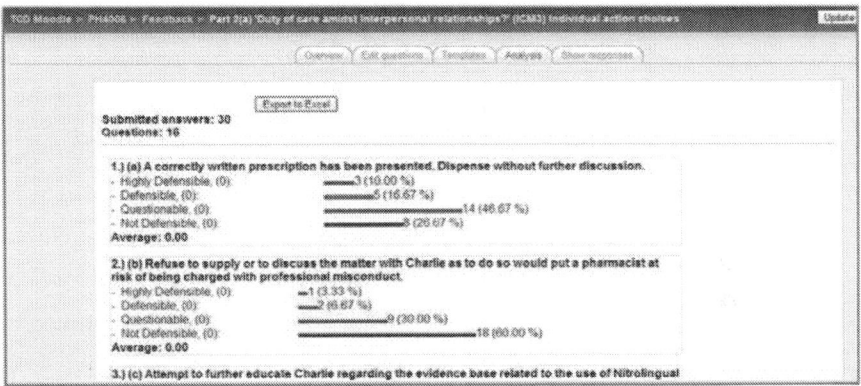

Figure 3. Part 2 (a): extract of group rating of the first two action options provided (Appendix 1).

- In the example shown in Figure 3, the student that suggested that (b) ("Refuse to supply or to discuss the matter with Charlie, as to do so would put a pharmacist at risk of being charged with professional misconduct") was highly defensible might reconsider, whereas the spread of response to (a) ("A correctly written prescription has been prescribed. Dispense without further discussion") is likely to generate interactive discussion and debate amongst the group.

- A further "teaching and learning" use to which this information was put is that when the three most and three least preferred options have been clarified, the six remaining action options are then extracted for use in a later workshop that seeks to focus on motivations and intentions that might underpin decision-making through dilemmas. These six action options, considered by this peer group to be neither "very defensible" nor "not defensible", are likely to demonstrate examples of behaviors that might be subjected to external scrutiny by supervisors or by a fitness-to-practice process after a dilemma "event", as proposed in the scenario. The subsequent workshop requires students to work in groups of three to post to the VLE what they believe might be the intention of a pharmacist that would chose each of these six action options.

- This scenario (Appendix 1) has, on occasion, been adapted to the hospital context by describing Celine as working in a small hospital on a Sunday morning. This type of variation supports students' understanding that dilemmas are likely to be encountered in any practice context. It also provides the flexibility to use the same scenario with two half-class groups to a useful effect, e.g., when applied to two half-class cohorts of fourth-year students, somewhat different outcomes regarding most and least preferred options were reached, and this provided a further opportunity to explore the concept that context does matter when reasoning through dilemma scenarios.

- ICM Part 3: group work.

- Teamwork, combined with group problem solving, is required to reach a group outcome as to the most "preferred" course of action. All students post individual choices online prior to being assigned to a group. Peer feedback is continuous as students debate and negotiate their way to a group decision. Criteria highlighted in the rubric (Appendix 3) guide the demonstration of the targeted competencies in a manner such that student behavior can be observed and assessed. VLE records provide evidence of the standard to which students have engaged in the process.

- The potential for the online use of ICMs to facilitate demonstration and assessment of targeted competencies, as outlined in the Core Competency Framework for pharmacists in Ireland (CCF) [19], is summarized in Figure 4.

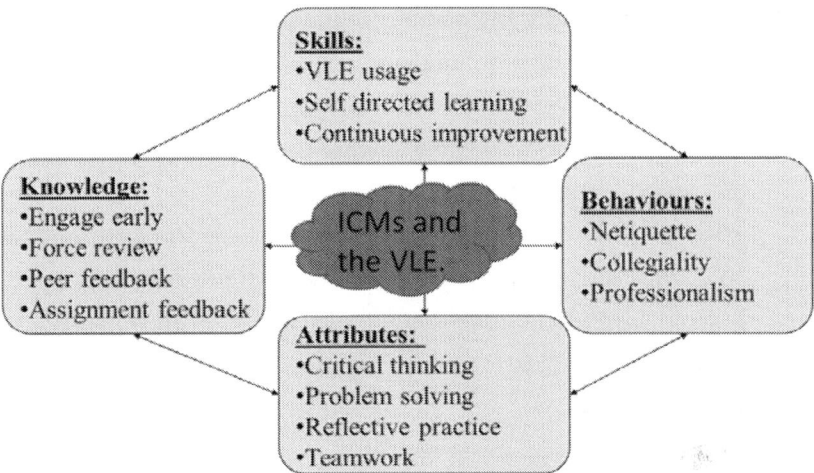

Figure 4. Intermediate concept measure (ICM) as a framework for demonstration and assessment of professional attributes [36]. VLE, virtual learning environment.

- The extent to which ICM Part 3 is completed online may be adapted to the stage in which the program and/or choices are made with respect to curriculum design. ICMs are introduced to first-year students in the final of a series of three workshops, and while Parts 1 and 2 are completed online, the group work is face-to-face with the potential to upload agreed upon group decisions online. Students must engage in the process in a manner considered satisfactory by the workshop leader, but no percentage or grade is allocated to the activity.

- Fourth-year undergraduate students, however, complete the process online over a 10-day period, guided by the rubric (Appendix 3), and the group work required to complete Part 3 of the ICM accounts for 7.5% of the overall module.

- (ii) Post graduate MPharm: online learning while interns undertake a 12-month placement under the direction of a tutor at a location in the Republic of Ireland (ROI).

- Pharmacy students in the ROI who have completed a primary degree in pharmacy (B.Sc.Pharm or B.Pharm) apply to enter the

National Pharmacy Internship Program (NPIP), which is delivered by the Royal College of Surgeons in Ireland (RCSI) on behalf of the Pharmaceutical Society of Ireland (PSI). Interns complete this year in approved training establishments under tutor supervision, and the author does not meet the interns face-to-face during the program. Interaction and assessment is based on three one-week cycles. Three questions, rather than the five presented in Appendix 1, were used in the independent review of the dilemmas scenario (referred to as Phase 1/Week 1), action options were presented in Phase 2/Week 2 without the subsequent provision of justification options, and the group sizes (Phase 3/Week 3) were up to seven rather than groups of up to five, with undergraduate students. The NPIP curriculum design included the development of a series of 17 podcasts, including one on principlism as a "tool to reason with", and podcasts were made available for download. An outline of the process is provided in Figure 5.

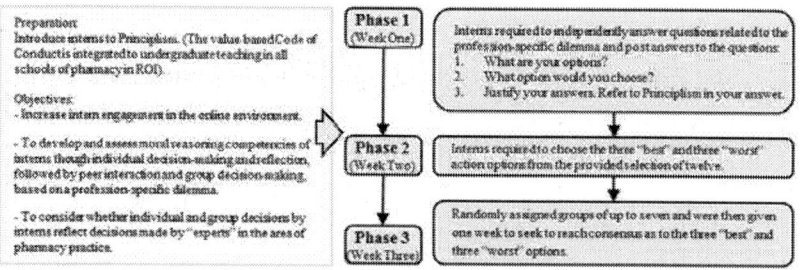

Figure 5. Professionalism and ethics dilemma review process, assignment of the National Pharmacy Internship Program (NPIP) [37]. ROI, Republic of Ireland.

- Multiple small groups engaged in teamwork, or team-based learning, can be observed and facilitated simultaneously by one academic, thereby accommodating the demands of large cohorts in geographically removed locations. Notwithstanding that significant staff time demands result from the support of multiple discussion fora and online feedback processes, the methodology nevertheless supports a means by which the expertise of a skilled "ethics education facilitator" can be made available to large cohorts of students, and geographically remote students can be supported as they develop competencies related to ethical reasoning. The objectives included that intern engagement in the online environment be increased, that a means by which moral reasoning competencies could be assessed in the context of the NPIP be developed and that a review of the outcomes quantify the extent to which individual and group decision-making by interns reflected "expert" opinion. Data gathered, directly from the VLE during the first three years of the NPIP highlights how the use of ICM methodology in the VLE provides a framework by which objective data related to intern performance and activity can be gathered. Intern downloading of podcasts, interactions online and intern and group performance, as determined by the grades achieved, were reviewed to provide the summary presented in Figure 6.

Phase 1/Week 1
- Student access to podcast on Principlism
- Principlism reference in review process.
- Independent review of dilemma (+CoC).

	2010 (n = 127)	2011 (n = 125)	2012 (n = 148)
Podcast (average 17)	98% (43%)	88% (49%)	83% (53%)
Min 3 Ps	92%	83%	66%
Score > 70%	98%	88%	96%

Phase 2/Week 2
- Reflection on action options provided.
- Alignment with peers/practitioners.
- Change options to reach group outcome.

	2010	2011	2012
Align Option 1	58%	52%	48%
Align Option 3	100%	82%	87%
% change to group	87%	96%	98%

Phase 3/Week 3
- Peer interaction: per person (group).
- Team alignment with practitioners.

	2010	2011	2012
Interaction/person (group)	3.5 (22)	5.3 (31)	4.4 (26)
Align Option 1	70%	62%	56%

Figure 6. Outcomes of a professionalism and ethics dilemma review assessment in the NPIP [38].

- The results provided in Figure 6 show that in, e.g., 2011, 88% of interns downloaded the podcast on principlism, while the average download rate across the 17 podcasts was lower at 49%.

Furthermore, 83% of interns referred to at least three of the four principles (autonomy, beneficence, non-maleficence and justice) introduced in that podcast [28], indicating that students had internalized the material on the podcast, and 88% of interns achieved 70% or more when completed assignments uploaded during Phase 1/Week 1 were graded. The scenario provided in Appendix 1 was used in 2011.

- When interns (2011) responded individually (Phase 2/Week 2), 52% of interns' most preferred action options aligned with collated expert opinion, showing that students recognized the norms of the profession in this regard, and 82% of interns included the option most preferred by the experts amongst their top three choices (Phase 3/Week 3). Most interns (96% in 2011) changed their ranking of the most and/or least preferred action options in order to agree with the group decision, thereby reflecting a requirement for debate and negotiation to complete the task.

3. SUMMARY

This article introduces the background to ICMs and presents the ICM as a tool that has potential to support professional education. It also shares the learning experienced by one pharmacist using ICMs in pharmacy education, provides an example of how a profession-specific ICM might be formatted, suggests how the methodology might be used in various ways in undergraduate and postgraduate pharmacy education and provides samples of measurables that may be incorporated into evaluation and assessment systems, both for educational interventions delivered face-to-face or partly or entirely online.

There are limitations to any inferences that can be taken from the use of ICMs to support the development, demonstration and assessment of professional attributes as proposed, including, but not limited to, the following:

- The examples outlined all relate to the ROI and cultural/legal variations, including legislation specific to the medication(s) referred to in the scenario, must be considered before deciding whether these might be generalizable to other jurisdictions.

- Profession-specific intermediate concepts central to other professions may differ from those for pharmacy and need to be considered in order to write appropriate ICMs for educational initiatives other than for pharmacy or for multidisciplinary group work.

- The technology itself raises concerns: (1) Assessment strategies need to evolve to manage the risk of plagiarism and impersonation. (2) Technology creates a different communicative space, with a permanent record of all interactions. Educators have a responsibility to seek to protect these incoming students from naivety in this regard. (3) The VLE must be adapted to accommodate the automation of teaching and learning where viable. Reservations regarding reduction in group size derive at least partly from the time pressures (on tutors/academics moderating and/or assessing group work) associated with those changes. (4) Strict cut-off times mean that there will inevitably be late-comers, and the accommodation of these participants, essential where assessment is involved, can be challenging [39]. (5) It can be challenging to accommodate "repeat" assessments for individuals when the focus is on group work.

- Feedback from the students [39] highlights that the tutors must pay particular attention to netiquette guidelines that prompt timely engagement by all group members, so that those engaging in the early stages do not become prematurely disheartened with the online team work process.

Several areas merit further consideration, including:

- The potential for the FCM to be used as an overarching approach to professional ethics education in pharmacy, wherein all four components would be specifically targeted.

- The establishment of a multisite collaborative investigation of the use of these methodologies in undergraduate pharmacy education.

- The adaptation of the teaching and assessment techniques outlined in this article to continuing professional development initiatives for practicing pharmacists.

- The review of how these methodologies might be adapted to incorporate interprofessional and interdisciplinary learning through the use of multidisciplinary online groups.

- The use of the VLE in online and blended learning programs merits further review as a means of supporting teaching and assessment methodologies aligned with competency-based assessment, such as those currently being introduced for pharmacy programs in the ROI [19]. The approaches used have the potential to support assessment challenges surrounding professional attributes, *i.e.*, in order to assess competencies, the related behaviors must first be demonstrated in a manner that can be externally observed. The use of the online environment facilitates the demonstration of professional attributes, such as teamwork, in a manner that can be observed and assessed, even where resources (e.g., tutor time) are relatively restricted.

ACKNOWLEDGMENTS

The authors acknowledge Muriel Bebeau for her thought-provoking conversations on professional ethics education and the four component model. Marek Radomski, Paul Gallagher and the students and colleagues involved in or supportive of the educational interventions referred to throughout this article are thanked.

AUTHOR CONTRIBUTIONS

Cicely Roche, Associate Professor in the practice of pharmacy, developed and delivered the educational material reviewed and wrote the initial draft of this article. Stephen Thoma, Professor in educational psychology, provided guidance and expertise in the Neo-Kohlbergian approach to moral development. Joy Wingfield, honorary Professor in pharmacy law and ethics, provided guidance and expertise in pharmacy-specific professional education. All authors read and approved the final manuscript.

REFERENCES

1. Thoma, S.J.; Bebeau, M.J.; Bolland, A. The role of moral judgment in context—Specific professional decision making. In *Getting Involved: Global Citizenship Development and Sources of Moral Values*; Sense Publishers: Amsterdam, The Netherlands, 2008; pp. 147–160.

2. Bebeau, M.J.; Thoma, S.J. "Intermediate" concepts and the connection to moral education. *Educ. Psycol. Rev.* **1999**, *11*, 343–360.

3. Bebeau, M.J. The defining issues test and the four component model: Contributions to professional education. *J. Moral Educ.* **2002**, *31*, 271–291.

4. Rest, J.; Narvaez, D.; Bebeau, M.J.; Thoma, S.J. *Post-conventional Moral Thinking: A Neo-Kohlbergian Approach*; Erlbaum: Mahwah, NJ, USA, 1999.

5. Rest, J.R.; Narvaez, D. *Moral Development in the Professions: Psychology and Applied Ethics*; Lawrence Erlbaum Associates: Hillsdale, NJ, USA, 1994.

6. Bebeau, M.J.; Monson, V.E. Guided by theory, grounded in evidence: A way forward for professional ethics education. In *Handbook on Moral and Character Education*; Nucci, L., Narvaez, D., Eds.; Routledge: New York, NY, USA, 2008.

7. Thoma, S.J.; Mahwah, N.J. Research on the defining issues test. In *Handbook of Moral Development*; Killen, M., Smetana, J.G., Eds.; Lawrence Erlbaum Associates: Mahwah, NJ, USA, 2006; pp. 67–92.

8. Roche, C.; Kelliher, F. Giving "best advice": Proposing a framework of community pharmacist professional judgement formation. *Pharmacy* **2014**, *2*, 1–12.

9. Chaar, B. Decisions, decisions: Ethical dilemmas in practice (or how to pass the "Red Face Test"). *Aust. Pharm.* **2006**,*25*, 444–449.

10. Chaar, B. Legislative change in Australian pharmacy—History in the making. *Aust. Pharm.* **2010**, *29*, 198–199.

11. Roche, C.; Kelliher, F. Exploring the patient consent process in community pharmacy practice. *J. Bus. Ethics* **2009**, *86*, 91–99.

12. Cooper, R.J.; Bissell, P.; Wingfield, J. "Islands" and the "Doctor's Tool": The ethical significance of isolation and subordination in UK community pharmacy. *Health* **2009**, *13*, 297–316.

13. Cooper, R.J.; Bissell, P.; Wingfield, J. Ethical decision-making, passivity and pharmacy. *J. Med. Ethics.* **2008**, *34*, 441–445.

14. Cooper, R.J.; Bissell, P.; Wingfield, J. Ethical, religious and factual beliefs about the supply of emergency hormonal contraception by UK community pharmacists. *J. Fam. Plan. Reprod. Health Care* **2008**, *34*, 47–50.

15. Cooper, R.J.; Bissell, P.; Wingfield, J. Dilemmas in dispensing, problems in practice? Ethical issues and law in UK community pharmacy. *Clin. Ethics* **2007**, *2*, 1–6.

16. Benson, A.; Cribb, A.; Barber, N. Understanding pharmacists' values: A qualitative study of ideals and dilemmas in UK pharmacy practice. *Soc. Sci. Med.* **2009**, *68*, 2223–2230.

17. Latif, D.A. The relationship between pharmacists' tenure in the community setting and moral reasoning. *J. Bus. Ethics***2001**, *31*, 131–141.

18. Latif, D.A. Cognitive moral development and pharmacy education. *Amer. J. Pharm. Educ.* **2000**, *64*, 451–454. Pharmaceutical Society of Ireland (PSI). Core Competency Framework for Pharmacists. 2009. Available online: http://www.thepsi.ie/gns/home.aspx (accessed on 31 December 2013).

19. Wingfield, J.; Bissell, P.; Anderson, C. The Scope of pharmacy ethics— An evaluation of the international research literature, 1990–2002. *Soc. Sci. Med.* **2004**, *58*, 2383–2396.

20. Roche, C. Ethical and legal issues in healthcare: Residential care: Pharmacist dilemmas and issue of "covert" medication. *Iran. Pharm. J.* 2010, 87, pp. 203–204. Available online: http:// www. thepsi. ie/tns/publications/irish-pharmacy-journal/ethics-articles.aspx (accessed on 31 December 2013).

21. Sanghavi, N. Pharmacy in a new age commercial environment. *Pharm. J.* **1995**, *255*, 615–618.

22. Szeinbach, S.L.; Barnes, J.H.; Summers, K.L.; Benjamin, B.F., III. The changing retail environment: Its influence on professionalism in chain and independently owned pharmacies. *J. Appl. Bus. Res.* **1994**, *11*, 5–15.

23. Swisher, L.L.; van Kessel, G.; Jones, M.; Beckstead, J.; Edwards, I. Evaluating moral reasoning outcomes in physical therapy ethics education: Stage, schema, phase and type. *Phys. Ther. Rev.* **2012**, *17*, 167–175.

24. Jones, D. A novel approach to business ethics training: Improving moral reasoning in just a few weeks. *J. Bus. Ethics* **2008**, *88*, 367–379.

25. Treleaven, L.; Voola, R. Integrating the development of graduate attributes through constructive alignment. *J. Market. Educ.* **2009**, *30*, 160–173.

26. Biggs, J. Constructing learning by aligning teaching: Constructive alignmen. In *Teaching for Quality Learning at University*, 2nd ed.; SRHE and Open University Press: Berkshire, UK, 2004; pp. 11–33. Beauchamp, T.L.; Childress, J.F. *Principles of Biomedical Ethics*; Oxford University Press: New York, NY, USA, 2009.

27. Pharmaceutical Society of Ireland (PSI). Code of Conduct for Pharmacists. Available online: http://www.thepsi.ie/ gns/ home. aspx (accessed on 31 December 2013).

28. Wingfield, J.; Badcott, D. *Pharmacy Ethics and Decision Making*; Pharmaceutical Press: London, UK, 2007.

29. Hew, K.F.; Cheung, W.S. Student facilitators' habits of mind and their influences on higher-levelknowledge construction occurrences in online discussions: A case study. *Innov. Educ. Teach. Intern.* **2011**, *48*, 275–285.

30. Sthapornnanon, N.; Sakulbumrungsil, R.; Theeraroungchaisri, A.; Watcharadamrongkun, S. Social constructivist learning environment in an online professional practice course. *Am. J. Pharm. Educ.*; 2009; 73, pp. 1–11. Available online: http://www.ncbi.nlm.nih. gov/ pmc/ articles/PMC2690880/ (accessed on 25 April 2014).

31. Huball, H.; Burt, H. An integrated approach to developing and implementing learning centred curricula. *Int. J. Acad. Dev.* **2004**, *1*, 51–65.

32. Hrastinski, S. A theory of online learning as online participation. *Comput. Educ.* **2009**, *52*, 78–82.

33. Feinberg, Jonathan. Wordle™. Available online: http:// www. wordle. net/ (accessed on 25 April 2014).

34. Roche, C. Technology Enhanced Pharmacy Education: Using the Virtual Learning Environment to Support the Development of Professional Attributes. Poster Presentation. In Proceedings of the International Pharmaceutical Federation (FIP) 73rd World Congress, Amsterdam, Netherlands, 3–8 October 2012.

35. Roche, C.; Gallagher, P. Developing moral reasoning skills in the virtual learning environment (VLE). Poster Presentation. In Proceedings of the Association for Medical Education in Europe (AMEE) Annual Conference, Glasgow, Scotland, 4–8 September 2010.

36. Roche, C.; Gallagher, P. Technology Enhanced Development of Moral Reasoning Competencies in Pharmacist Interns: Stimulating Their Engagement in Dilemma Review and Resolution in the Online Environment. Oral Presentation. In Proceedings of the International

Pharmaceutical Federation (FIP) 73rd World Congress, Amsterdam, Netherlands, 3–8 October 2012.

37. Roche, C. Formative Assessment for "Graduate Attributes": Technology-Enhanced Learning in the First Semester. Proceedings of the Edulearn12 Conference Proceedings, Barcelona, Spain, 1–3 July 2012; Available online: http://library.iated. org/view/ ROCHE2012FOR (accessed on 23 February 2014).

38. Trinity College Dublin School of Pharmacy and Pharmaceutical Sciences (B.Sc.Pharm) Degree Course Student Handbook 2012–2013. Available online: http://pharmacy.tcd.ie./pdf/Student%20Handbook%20%202012%2013%20FINAL.pdf (accessed on 25 April 2014).

39. Dublin City University. Using marking schemes/rubrics—DCU. 2014. Available online: http://pharmacy.tcd.ie./pdf/Student%20Handbook%20%202012%2013%20FINAL.pdf (accessed on 25 April 2014).

CHAPTER 5

Implementation of E-Learning System: Findings and Lessons Learned

Namsraidorj Munkhtsetseg, Sambuu Uyanga

Department of Information System, School of Mathematics and Computer Science,
National University of Mongolia, Ulaanbaatar, Mongolia

ABSTRACT

This paper describes the implementation of the e-learning system at the School of Mathematics and Computer Science, National University of Mongolia. The paper includes in-house development of Edunet 1.0 e-learning system, comparative analysis on LMS, evaluation methodology, selection of e-learning systems, and comparative analysis on implementation of Edunet, Moodle and Canvas systems.

KEYWORDS

E-Learning System; Moodle; Canvas; LMS; Qualitative Weight and Sum Approach; National University of Mongolia

1. INTRODUCTION

The National University of Mongolia (NUM) is the country's oldest and only comprehensive university and a leading center of science, education and culture. It has more than 30 schools, institutions and research centers. There are more than 10,000 students in 16 branch schools. The School of Mathematics and Computer Science (SMCS) of the National University of Mongolia has three branches: Theoretical Mathematics, Application Mathematics and Information Technology.

The School of Mathematics and Computer Science (SMCS) is planning to build infrastructure model for elearning. This paper addresses selection and evaluation of most appropriate e-learning system within the above objective.

Building the infrastructure for online learning requires that many factors be considered, so it is difficult to provide a straight-forward checklist or recipe to follow. All educational endeavors are systems, made up of various interconnected components [1].

Our infrastructure model for e-learning system consists of following interconnected components:

• E-learning system;

• University Management Information System;

• E-library system;

• E-content development;

• Other services to students.

There are two development aspects:

1) Waterfall development [2]. To develop complex information system, which includes all sub-systems;

2) Agile development. To develop system by developing sub-systems separately and integrating them.

Usually developers use a waterfall method when developing e-learning. The most used one is the ADDIE model, where development has five phases: analysis, design, development, implementation and evaluation. It worked for years but it takes a long time to go through all the phases, not really suited for on demand responses. We need to get faster and more iterative [3].

The iterative nature of agile development means features are delivered incrementally, enabling some benefits to be realized early as the product continues to develop. A key principle of agile development is that testing is integrated throughout the lifecycle, enabling regular inspection of the working product as it develops [3].

Therefore we selected Agile development method for building infrastructure model for e-learning system.

The SMCS has been implementing an open source e-learning system based on international standard since 2009. During this period we have implemented several systems [4].

This work addresses development and implementation of systems; comparisons of LMSs; survey on system usage are also presented. The first section describes an in-house development of e-learning system. The second section addresses comparative analysis on LMSs using Edutools; [5] the third—Comparison on adaptation of open source e-learning systems by using evaluation Methodology: Qualitative Weight and Sum Approach [6]; the fourth—Survey on System Usage. The last section comments on the advantages and disadvantages on development and implementation of e-learning system.

2. IN-HOUSE DEVELOPMENT OF E-LEARNING SYSTEM

We developed e-learning system named Edunet 1.0 [4] on Rails framework based on Ruby programming language [7] with MVC [8] architecture. The system has following modules or functionalities:

- Message: This module supports information flows: Education department-lecturer, lecturer-lecturer and lecturer-student. Also file attachment is available.

- Course: This module allows students to receive all information and lecture materials of specific course.

- File: This module allows to lecturer to upload course related files to the system and to students to download course materials.

- Homework: This module allows the Education department and lecturer to receive tasks and homework in file format within pre-defined period.

- Quiz: Quiz module allows the lecturer to set quiz tests, to set a time period for testing and to export to MS excel file. Quizzes can allow multiple attempts.

- Journal: This module allows the lecturer to insert student's entry, activity or participation. The marks for quizzes will be inserted automatically to the journal.

- Discussion: This module provides a simple communication method between lecture and students. A student can open a dialogue with a lecturer and ask questions.

This system is important in terms of supporting learning activities and helping to conduct training. However this system does not meet all functionalities of modern learning systems such as discussion, content development and electronic presentation of course materials etc. Thus, the users requested updates.

Due to the lack of human resources in the development team, it was impossible to improve the content development and other technical issues in the system. Therefore, we have decided to implement another e-learning system [6].

The following problems occurred during the two years implemetation of the Edunet 1.0 [4] system:

1. Development of new function, module or application according to each new requirement. Duration of development takes usually 1 - 2 months;

2. Lecturers send lecture materials to students directly;

3. Quiz module supports only few simple types of questions;

4. Impossible to use course materials in the next academic year or backup course materials;

5. Impossible to analyze students activity or participation;

6. Lack of key e-learning modules such as team management, virtual classroom etc.

3. COMPARATIVE ANALYSIS ON LMS

For comparing LMS products, we used www.edutools.info tools [5]. The comparative analysis on key modules of LMSs, such as Blackboard [9], Angel [10], JoomlaLMS [11], Moodle [12], and ATutor [13] described in **Table 1**, where "+" marks availability of current function. For example, five "+" for file exchange module of Black-Board LMS and one "+" for Moodle system. It means the Blackboard has a five different method of file exchange and Moodle has only one.

Our study shows that commercial systems have more functionalities than open source free systems. But open source systems have most of key necessary functionalities and possibility for future development according to the user needs and requirement.

Therefore based on our studies we decided to implement Moodle [12] open source system.

4. COMPARISON ON ADAPTATION OF OPEN SOURCE E-LEARNING SYSTEMS BY USING EVALUATION METHODOLOGY: QUALITATIVE WEIGHT AND SUM APPROACH

One of key issues to consider when developing and implementing e-learning systems is the adaptation of the current system to the cognitive learning characteristics of the students. Implementation of the adaptation is not a simple process, since it implies the study and conjunction of technical and pedagogical issues [14].

Therefore we evaluated an adaptation of systems based on evaluation study of open source e-learning platforms/ Virtual Learning Environment [15]. The most adaptable system is selected on results of evaluation on adaptaion systems using Qualitative Weight and Sum Approach method [6].

There are two main points on evaluation of e-learning systems:

• Selection of modules to be evaluated;

• Selection of evaluation method.

Evaluation was carried out by using the Qualitative Weight and Sum Approach method according to the IEEE LTSA [6] reference model. The qualitative weight and sum (QWS) approach is a well-established approach for the evaluation of software products. It establishes and weights a list of criteria. QWS is based on the use of symbols. There are six qualitative levels of importance for the weights, frequently symbols are used: * = extremely valuable, # = very valuable, + = valuable, | = marginally valuable and 0 = not valuable.

We considered following criterias defined by P. Baumgartner and H. Häfele [16]:

• Support of dynamic communication;

• Sustainability of development;

• Good documentation of the system.

4. Comparison on Adaptation of Open Source E-Learning Systems by
Using Evaluation Methodology: Qualitative Weight and Sum

117

Table 1. Comparative analysis on LMS.

Product Name	Blackboard [9]	ANGEL [10]	Joomla LMS [11]	Moodle [12]	ATutor [13]
Developer Name	Blackboard	ANGEL	E-Learning Force Inc.	Moodle	ATRC Uni of Toronto
Communication Tools					
File Exchange	+++++	++++	+++	+	++++
Online Journal/Notes	+++++	+++	++		+
Real-Time Chat	++++++	++++++++	+++++	+++++	+++
Whiteboard	++++++	+++++++	+++++	+	+
Administration Tools					
Authentication	++++++++	++++++++	++++++	++++++++	++++
Course Authorization	+++++	++	++++	++++	++++
Registration Integration	+++++++	+++++++	++++	+++++++	+++
Hosted Services	++	++	++	++	++
Content Development Tools					
Accessibility Compliance	++++	++++	++	++	++++
Content Sharing/Reuse	+++++	+++++	++	++	++++
Course Templates	+++++	+++++	++++	+++	++
Customized Look and Feel	++++++	++++++	++++	+++++	++
Instructional Design Tools	++++	++++	+++	+++	+++
Instructional Standards Compliance	+++SCORM 2004	++++SCORM 2004	++	+++	+++
Company Details/Licensing					
Costs/Licensing	Commercial	Commercial	Commercial	GPL	GPL
Open Source	No	No	No	Open Source	Open Source

The following modules of the ATutor, Dokeos, ILIAS, Moodle, OpenUSS, Sakai, and Canvas systems have been examined:

- Communication tools;

- Learning objects;

- Management of user data;

- Usability;

- Adaptation;

- Technical aspects;

- Administration;

- Course management.

After defining above modules we divided these modules into sub-modules and carried out weight of each sub-module. Maximum value is a maximum value of current criteria. List of modules and sub-modules by each system are presented in **Table 2**, where red color is best one. **Table 2** shows Canvas-five, Moodle-two and ILIASone best result. Summarized evaluation is presented in **Table 3.Figure 1** shows that weights of Canvas [17] and Moodle systems are higher than others.

We started to implement Moodle system since 2009- 2010 academic years and developed 20 course materials to the system. We are still using this system in conjunction with other systems. We also decided to implement the Canvas Learning Management System that was selected because of its:

- clean and user-friendly interface;

- rich collaboration and discussion tools;

- ability to embed multi-media and web-based resources;

- integration with communication tools such as Facebook, GoogleDocs, Skype and Google Apps for Education, and compatibility with other existing teaching technologies [15];

- great course management and organization tools including a calendar that integrates and displays due dates and academic activities for students;

- well-designed rubrics and grading tools.

Within the implementation of the system, we:

1. Implemented vitual server at http://lesson.num.edu. mn and uploaded more than 200 hours lecture materials;

2. Carried out the localization of the Canvas system to Mongolian language;

3. Uploaded localized system to the server conducted training on system usage to all lecturers.

4. Comparison on Adaptation of Open Source E-Learning Systems by Using Evaluation Methodology: Qualitative Weight and Sum

Table 2. Evaluation results of e-learning platforms for each subcategory.

Sub Categories	Communication Tools							Learning Objects							Management of User Data		Usability				Adaptation				Technical Aspects				Administration				Course Management	
	Forum	Chat	Mail messages	Announcements Announcements	Conferences	Collaboration	Synchronous & Asynch Tools	Tests	Learning Material	Exercises	Other Creatable LOs	Importable LOs	Tracking	Statistics	Identification of Online Users	Personal User Profile	User-Friendliness	Support	Documentation	Assistance	Adaptability	Personalization	Extensibility	Adaptivity	Standards	System Requirements	Security	Scalability	User Management	Authorization Management	Installation of the Platform	Administration of Courses	Assessment of Tests	Organization of Course Objects
Maximum Values	*	*	|	+	+	+	*	*	*	#	+	*	*	*	+	+	#	#	#	+	+	*	#	*	*	#	+	*	*	+	#	*	|	+	#	#
Atutor	|	#	|	|	0	0	*	|	*	0	+	*	*	+	|	|	+	|	+	+	|	#	#	|	+	*	+	0	0	0	|	|	|	|	#
Dokeos	+	*	0	|	+	0	*	*	*	0	+	*	+	|	0	|	+	#	+	|	|	0	*	+	+	+	0	0	#	0	|	|	|	|	#
ILIAS	+	*	|	0	0	0	*	*	|	0	+	*	|	|	+	+	|	|	+	0	+	#	*	0	#	*	+	*	0	#	*	|	+	*	+	+
Moodle	*	*	0	+	0	+	*	*	*	#	+	*	*	|	+	+	+	#	+	+	#	+	*	|	#	+	+	+	+	|	|	|	|	|
OpenUSS	#	*	0	+	0	|	*	0	|	0	+	#	0	0	+	+	+	+	|	+	#	#	#	0	0	|	+	0	0	0	0	|	#
Sakai	#	*	0	|	0	0	*	0	*	#	|	*	*	0	|	|	#	|	|	0	0	0	*	0	0	+	+	+	0	+	|	+	0	0
Canvas	*	*	|	+	+	+	*	*	+	#	+	*	*	+	0	#	#	|	|	#	#	|	|	#	*	+	|	*	+	|	*	0	+	+	#

Table 3. Summarized evaluation by using method qualitative weight and sum approach.

	0 = not valuable	| = marginally valuable	. = valuable	# = very valuable	* = extremely valuable
Moodle	2	8	11	5	8
ILIAS	7	7	10	3	7
Dokeos	8	8	9	3	6
Atutor	6	13	7	4	4
OpenUSS	11	6	9	6	2
Sakai	13	7	5	3	6
Canvas	2	8	9	7	8

According our survey and analysis, learning management systems should have following functionalities [18]:

- Logon/authentication with high security;

- Ability to configure according to user requirements;

- Ability to integrate student activities;

- Curriculum development with course specifics and selected learning methods;

- Course management;

- Student enrollment;

- Communication support (social networking, discussion forms, live chat etc.);

- Support of SCORM 2004.

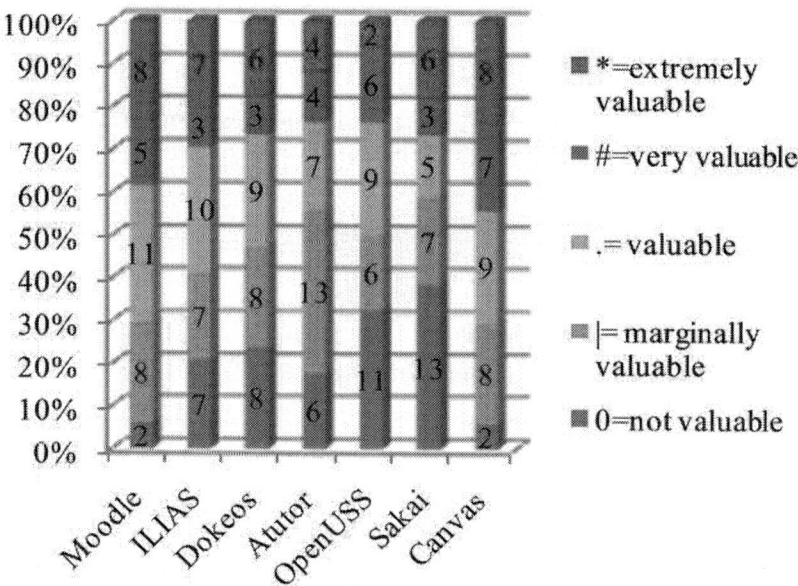

Figure 1. Summarized evaluation by using method qualitative weight and sum approach.

Table 4 shows that our developed Edunet system has only three of the 12 above mentioned functions.

5. SURVEY ON SYSTEM USAGE

We conducted a survey on those systems in 2009-2010 and 2010-2011 academic years. The survey covers total of 1426 students in 2009-2010, 756 students in 2010- 2011, and 670 students in 2011-2012 academic years. Survey results are presented in Tables 5 and 6.

Table 4. Comparison of three implemented systems.

No	Product Name	Edunet (In-House)	Moodle (Open Source)	Canvas (Open Source)
	Developer Name	NUM of SMCS	Moodle	Instructure.com
1	File Exchange	+	+	+
2	Online Journal/Notes	-	+	+
3	Real-Time Chat	+	+	+
4	Whiteboard	-	-	+
5	Authentication	+	+	+
6	Accessibility Compliance	-	-	+
7	Content Sharing/Reuse	-	-	+
8	Course Templates	-	+	-
9	Customized Look and Feel	-	+	+
10	Instructional Design Tools	-	+	+
11	Instructional Standards Compliance (SCORM 2004)	-	+	+
12	Costs/Licensing	Custom	Open Source	Open Source
	Summary	3	9	11

Based on our survey, we compared system normal use with all three systems using following formula:

$$X = Y * 100/Z$$

where, X-percentage of quality of current indicator, Ynumber of responses on Moodle and Canvas systems, Ztotal number of students participated in survey.

Final result is presented in **Table 7**.

Figure 2 shows that most indicators were unsatisfactory for the in-house Edunet system.

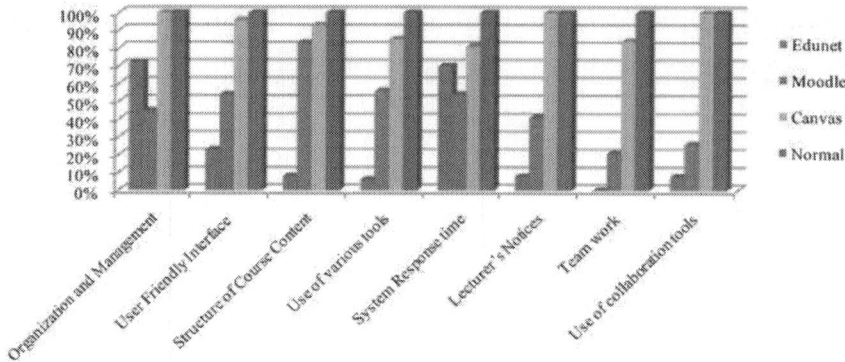

Figure 2. Comparison on system usage.

The system response time for Edunet system was higher than Moodle system because of the functionalities of the system are less than Moodle and the system doesn't require additional libraries. Because of its use of web services for team works and collaboration activities, related indicators for the Moodle were unsatisfactory in our survey. The most indicators were satisfactory for the Canvas system in development mode and approximate to the normal level. System response time can be higher in production mode than development mode.

6. CONCLUSIONS

The success of e-learning in tertiary education may be attributed to the following factors:

- Sustainable finding;

- Total commitment and support from top management;

- Participation, cooperation and support from other major universities and IT communities;

- Sufficient development and support staff with advanced technical skills;

- Strong technical support;

- Expertise in instructional design.

Considering the rapid development of the ICT, especially that of the educational technologies, networks and software, design and implementation of an e-learning system, the customized system in-house by the institute itself is not best option. Because, our lessons and practices show that, the in-house development of learning system by academic institutions (especially in developing countries) may experience following disadvantages:

- Development and implementation costs are high;

- The development team is not fully familiar with the standards of e-learning systems for which in most cases, some training classes are held to educate the team.

- System development demands a close collaboration between the IT professionals and the educational experts.

- If the development team's output is low, the actual overall expenses for this approach would be higher than other solutions [8].

Table 5. Survey results: 2010-2011 academic years.

	Indicators/Systems	Edunet		Moodle	
		Satisfied	Not Satisfied	Satisfied	Not Satisfied
1	Organization and Management	71.88%	28.13%	45.31%	54.69%
2	User Friendly Interface	23.44%	76.56%	54.69%	45.31%
3	Structure of Course Content	7.81%	92.19%	83.59%	16.41%
4	Use of Various Tools	6.25%	93.75%	86.72%	13.28%
5	System Response Time	70.31%	29.69%	54.69%	45.31%
6	Lecturer's Notices	7.81%	92.19%	40.63%	59.38%
7	Team Work	0.00%	100.00%	0.00%	100.00%
8	Use of Collaboration Tools	7.81%	92.19%	26.56%	73.44%

The SMCS uses SISI University information management system. Current open source e-learning systems can receive users and course related information from SISI system using LDAP server in real-time.

Table 6. Survey results: 2011-2012 academic years.

	Indicators/systems	Moodle		Canvas	
		Satisfied	Not Satisfied	Satisfied	Not Satisfied
1	Organization and Management	24.71%	76.47%	100.00%	0.00%
2	User Friendly Interface	30.59%	69.41%	95.29%	4.71%
3	Structure of Course Content	70.59%	29.41%	92.94%	7.06%
4	Use of Various Tools	27.06%	76.47%	84.71%	15.29%
5	System Response Time	58.82%	29.41%	81.18%	18.82%
6	Lecturer's Notices	38.82%	61.18%	100.00%	0.00%
7	Team Work	23.53%	76.47%	94.12%	5.88%
8	Use of Collaboration Tools	23.53%	76.47%	100.00%	0.00%

Table 7. Comparison on system usage.

No	Indicators	Edunet	Moodle	Canvas	Normal
1	Organization and Management	72%	45%	100%	100%
2	User Friendly Interface	23%	54%	96%	100%
3	Structure of Course Content	8%	83%	93%	100%
4	Use of Various Tools	6%	86%	85%	100%
5	System Response Time	70%	54%	81%	100%
6	Lecturer's Notices	8%	41%	100%	100%
7	Team Work	0%	0%	94%	100%
8	Use of Collaboration Tools	8%	26%	100%	100%

We are planning to conduct online trainings using the Moodle and Canvas systems in future, and started to implement these systems, which cover following activities:

1. Conduct application trainings of these systems for lecturers and students;

2. Conduct trainings for system administrators;

3. Develop user manual of the Moodle and Canvas systems in Mongolian language;

4. Conduct trainings on development of e-courses for lecturers. These trainings will cover applications such as eXe, Courselab, Articulate, Ispring, Adobe Flash which support Scorm standard;

5. Import data of all system users and courses from SISI system.

REFERENCES

1. Terry Anderson & Fathi Elloumi, "Theory and Practice of Online Learning," 2010.

2. http://en.wikipedia.org/wiki/Waterfall_model

3. http://www.agiledevelopment.org/

4. NUM, SMCS in house system

5. http://www.edutools.info

6. S. Graf and B. List, "An Evaluation of Open Source E-Learning Platforms Stressing Adaptation Issues," Presented at ICALT 2005.

7. Ruby. http://rubyonrails.org

8. http://en.wikipedia.org/wiki/Model viewcontroller

9. http://www.blackboard.com

10. http://www.angellearning.com

11. http://www.joomlalms.com

12. http://moodle.org

13. http://atutor.ca

14. S. Graf and B. List, "An Evaluation of Open Source ELearning Platforms Stressing Adaptation Issues," Women's Postgraduate College of Internet Technologies, Vienna University of Technology, Vienna, 2002.

15. ISO/IEC 19796-3, "Information Technology—Learning, Education and Training—Quality Management, Assurance and Metrics—Part 3: Reference Methods and Metrics," 2009.

16. P. Baumgartner, H. Häfele, and K. Maier-Häfele, "ELearning Praxishandbuch —Auswahl von Lernplattformen."

17. http://instructure.com

18. H. Wharekure and K. Aoteareo, "Technical Evaluation of Selected Learning Management Systems," 2004.

CHAPTER 6

Efficacy and Usability of an E-Learning Program for Fostering Qualified Disease Management Nurses

Kana Kazawa[1], Michiko Moriyama[1], Michiyo Oka[2], Satsuki Takahashi[3], Madoka Kawai[4], Masumi Nakano[5]

[1]Chronic Disease Management Project Research Center, Institute of Biomedical and Health Sciences, Hiroshima University, Hiroshima, Japan
[2]Graduate School of Health Sciences, Gunma University, Gunma, Japan
[3]School of Nursing, Gunma Prefectural College of Health Sciences, Gunma, Japan
[4]Division of Nursing Science, Institute of Biomedical and Health Sciences, Hiroshima University, Hiroshima, Japan
[5]Hiroshima City Asa Hospital, Hiroshima, Japan

ABSTRACT

In order to train nurses to perform disease management and telenursing, we developed an e- learning education program, and assessed the efficacy. A single-group pre-test and post-test design was used. Nurses who worked at a medical institution or a disease management company were included, and the duration of the program was set 2 months. We developed the program so that it could grow attitude and improve knowledge and skills in disease management and patient education. Of 55 subjects, 48 who completed the program were analyzed. After the program, subjects increased knowledge and interests in disease management and patient education. Almost of the subjects answered that e-learning was a good learning method. Our program was effective at enhancing subject's interests in disease management and patient education, and considered to improve their skills in the future.

KEYWORDS

E-Learning, Education Program, Disease Management, Telenursing

1. INTRODUCTION

With the change of disease composition, the healthcare system in the world has been required to change the targets from infectious diseases to non-infectious diseases (chronic diseases). However, since the structural reform has not caught up the change, there are unmet needs for chronic disease care, resulting in disease worsening and medical expense increase [1] - [3] . The situation is the same in Japan, and to solve this problem, in 2013, the government obligated all health insurers to carry out a data-health project aimed at improving health in their insured people and keeping reasonable medical expense [4] . Before this motion, we put into practice a new model that provided chronic disease care based on the disease management methodology in Japan. These programs were effective at preventing disease progression in patients with chronic illnesses and may contribute to the containment of medical costs [5] - [11] . Disease management was an outcome management method that was proposed early by the Disease Management Association of America, in order to properly allocate limited medical resources and provide high-quality healthcare with containing medical expense [12] . Disease management includes the following factors: 1) identification of a target population, 2) provision of standardized medical treatment and patient education based on evidence-based clinical practice guidelines, 3) interprofessional collaboration, 4) analysis and evaluation of outcomes and processes, 5) feedback to health care professionals and patients [13]. In disease management, we considered that nurses were the best players because they were good at arranging with professionals and services and specialized for patient education. Therefore, we have fostered disease management nurses (DM nurses) who perform disease management. Furthermore, we also considered that in order to provide care widely, the nurses should obtain the ability of practical telenursing that allowed them to communicate with people in remote areas and contain healthcare expense. Therefore, we set up a program for fostering disease management nurses using electronic learning (e-learning) on disease management and patient education for patients with a chronic disease (e-education for DM nurses). We developed our program using e-learning, because we considered that e-learning could be a good way for nurses to learn at their own pace at workplace or home and that it could ensure the number of disease management nurses nationwide and guarantee their quality. Regarding the efficacy of e-learning for students, several articles reporting e-learning education used in nurse training faculties and they mentioned that e-learning was not inferior to the traditional face-to-face education in improving subject's knowledge and their practical ability as well as their satisfaction [14] [15] . Also, for the registered nurses, there are many articles reporting that education using e-learning resulted in improvement of subject's knowledge and skills [16] - [18] . These reports used themes such as diseases and nursing skills, and there has been no study on the theme of acquiring comprehensive ability including disease management and patient education. In addition, although it has been reported that patient education that is structured based on theories and evidences can improve outcome [19] . However, no study has shown the process of developing e-learning on improving the ability of patient education, and incorporated theories.

Therefore, in this study, we developed the e-learning education program on nurses who implemented disease management in order to foster comprehensive ability integrating knowledge and skills of disease management, patient education, and behavior modification. Then, we assessed the efficacy and the operability. If we can assure the quality of disease management nurses above a certain standard, it would guarantee the quality of disease management project in Japan.

2. E-EDUCATION FOR DM NURSES

2.1. Program Structure, Themes, and Goal Setting

The program structure is shown in Figure 1. Based on the classification of educational objectives proposed by Bloom et al. [20] , we incorporated three dimensions of attitude, knowledge and phsychomotor skills as the frameworks. The goal of our program was to "train generalist nurses to perform chronic disease management and telenursing" and generalist nurses were defined as nurses who had graduated from a basic nursing education course (professional school or university undergraduate education) and had clinical experience.

For the program structure, we selected outlines of disease management (including health assessment/nursing process), knowledge about characteristics of chronic diseases that are keys for disease management, care coordination, communication and quality management as the 5 themes. As "disease management", we included items such as definition and structure of disease management, program-developing process, evidence based medicine, patient education (including cognitive behavior therapy), and care triangulation (Emic: subjective approach taking into account Illness Narrative and Etic: objective approach taking into account data based on evidence). For the developing process, we built it up to follow the same course as nursing process that goes thorough comprehensive health assessment on patient's disease and risk factors from the physical, mental and social aspects, intervention, evaluation and feedback. For "knowledge of disease characteristics", we included diseases with a high incidence, i.e. metabolic syndrome, diabetes, diabetic nephropathy, ischemic heart disease, heart failure and stroke, and constructed the contents for the disease, diagnosis, standard treatment, medical care management and management goals and patient education, based on practice guidelines [21] -[25] . Furthermore, we included detailed contents to support acquiring self-management skills. For "care coordination", we defined that it was to assess needs of medical resources, support the patient and their family's decision making, and coordinate multidisciplinary collaboration. Therefore, in addition to skill of stratification, we educated on methods to assess medical resources and create medical and local collaboration, methods and skills of care coordination, and methods to support families as a care unit. For "communication", we educated on skills required for patient education, communication with physicians and telenursing. As skills required for patient education, the motivation interview methods, coaching, cognitive behavior therapy and behavior change skills were included [26] [27] . For "quality management", knowledge and skills required as research nurses (mainly, data monitoring, data management and personal data protection) and utilization of IT systems were included. Finally, as a summary of these lectures, we used videos illustrating physical examination and patient education techniques, such as the techniques for actual interviewing and telenursing (role play with simulated patients), for motivation interview methods, for management of diseases.

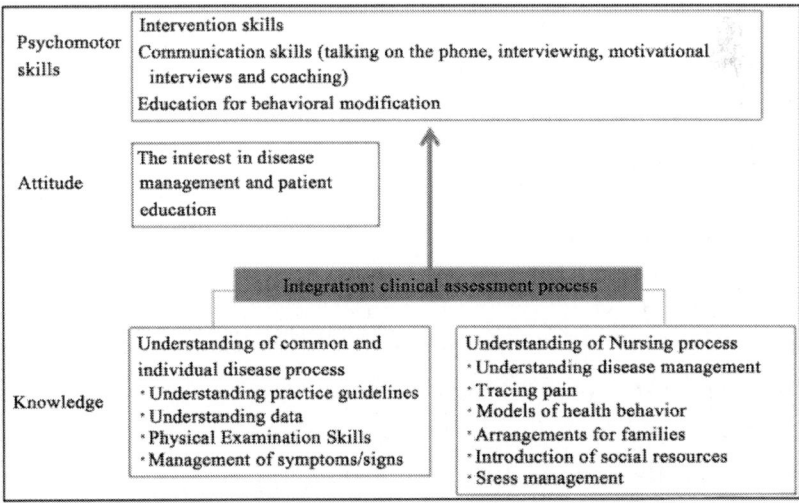

Figure 1. The structure of education program.

2.2. Development of E-Learning

Characteristically, e-learning is a way to learn subjectively, individually and repeatedly at learner's own pace, with enjoyment and easy understanding using senses of vision and hearing as well as a way to acquire the decision-making ability by following thinking process of experts in special fields [28] . We prepared lectures of 20 to 80 minutes for each theme and if the duration was longer than 20 minute, we divided it by 20 minutes, so that subjects could learn easily and repeatedly at their own pace.

For each theme, physicians and nurses specialized in the field, nurses specialized for managing chronic diseases, nurses certified for motivation interview, specialists for behavior change, physical therapists for exercise therapy and nutritionists for diet therapy performed the lecture. In the part of role play, nurses specialized for chronic disease management actually demonstrated. The screen layout of e-learning consisted of movies showing teachers giving their lecture and learning materials simultaneously.

2.3. Evaluation Tests

For individual themes, we set 10 yes-no questions based on the lecture to evaluate acquisition levels of learning.

3. METHODS

3.1. Subjects

Nurses who worked at an acute-care hospital or a disease management company, and agreed to participate in the study. Clinical experience should be at least 3 years and those who were during the postpartum period or the childcare leave could be included. We did not consider the specialized area of working and the level of basic education (university (undergraduate, graduate) or professional school). Exclusion criteria were assistant nurses, midwifes, nurses in manager positions (director of nursing, assistant director, ward manager).

3.2. Period of Analysis

April-December, 2014.

3.3. Research Design

A single-group pre-test and post-test design was used.

3.4. Primary and Secondary Endpoint

Evaluation indicators were set with reference to the classification of education targets [20] . In the classification, it is considered that most behaviors that arise from learning results involve 3 domains, i.e. "the cognitive domain" associated with acquiring knowledge, understanding and developing intellectual abilities, "the emotional domain" associated with developing interests, attitude and sense of value as well as emotion, proper judgment and adaptability, and "the psychomotor domain" associated with technical skills and motor skills. In addition, it is also considered that the emotional domain is corresponding to results from developing the cognitive domain. Based on this, learning effects were measured for the 3 domains, while we operationally defined the cognitive domain as "subjective evaluation on acquiring knowledge about disease management/patient education", the emotional domain as "interests in disease management and patient education" and the psychomotor domain as "acquiring skills in patient education" and used them as primary endpoints for education effects. The scales to measure achieved level of each domain were prepared as follows by researchers using Likert scale (Table 1).

Secondary endpoints were test scores evaluating acquired knowledge of each e-learning theme. For each subject, the average score was calculated for each theme.

In addition, for the program operability, subjects were asked to evaluate 1) the duration of e-learning (20- minutes sessions), 2) the numbers of themes and subthemes, 3) easiness of taking the program, 4) acceptance as a learning tool, in likert scale of 3 or 4 grades and also they could comment voluntary for the items.

3.5. Data Collection

Scores for the emotional domain, the cognitive domain and e-learning tests were obtained from subjects. The score for the psychomotor domain was obtained by research assistants who evaluated skills of subjects performing patient education on an actual setting. The research assistants only performed the evaluation and were not involved in analysis. As the schedule for data collection, scores for the emotional domain and the cognitive domain were collected at 2 time points before and after the program. Because we considered that it needed a lot of time to learn skills for disease management and patient education, scores for the psychomotor domain were collected at 1 time point after the program. Scores of e-learning tests were obtained when the subjects answered tests before and after completing each theme. Evaluation of the program was obtained at the end of the program.

3.6. Recruitment and Registration Procedures

Nurse managers who agreed to support this study introduced nurses who met the eligibility criteria. The researchers explained them about the objectives and contents of this program orally and in written form, after that obtained a written consent.

3.7. Program Implementation

We gave URL of the e-learning education program to subjects who agreed, and explained how to take the program and that they should complete the program within about 2 months from the enrolled day. We also told that they could take the program at any time and any place they wanted. In addition, from the view point of protecting intellectual property, we obtained a written consent to promise proper handling of the program URL.

3.8. Method of Analysis

To compare indicators over time, we used the t-test or the Wilcoxon signed-rank test, and the significance level was set at $p < 0.05$. Comments answered voluntarily were summarized inductively.

Table 1. Scales for evaluation of the emotional/cognitive/psychomotor domains.

Domain	Item	Point	Item number	Score range
Emotional domain (interest)	Interest in disease management Challenges for disease management Disease management is a skill that helps my career as a nurse Education for noncompliant/non-adherent patients Confidence to change behavior of noncompliant/non-adherent patients	1: not interested - 4: interested	5	5 - 20
Cognitive domain (knowledge)	Disease management How to apply evidence-based medicine to clinical practice Pathophysiology of diseases Treatments for diseases Examinations for diseases Self-management education Outline of care coordination Care coordination: stratification skills Care coordination: healthcare collaboration/local cooperation Care coordination: necessary healthcare resources Patient education Cognitive behavioral therapy/cognitive behavior theory Study method: study design, data management	1: not understood - 4: understood	13	13 - 52
Psychomotor domain (skills)	Health assessment (incl. physical examination): 18 items Patient education in self-management for behavioral change: 11 items Tele-communication skills: 7 items Communication with physicians: 5 items Attitude and manner as a professional: 5 items	1: not achieved - 4: achieved	46	46 - 184

3.9. Ethical Considerations

Approval was obtained from the ethics committees of Hiroshima University. The researchers explained to subjects about the purpose of this study, learning content, data collection, confidentiality, how the results would be published, and that they would be free to withdraw. Subjects gave a written consent.

4. RESULTS

4.1. Outlines of Subjects

Among 55 nurses who agreed to participate in the study, 48 completed the e-education (the completion rate: 87.3%). Seven nurses dropped out the e-education because they were too busy with their work or care giving.

As basic characteristics of subjects, there were 45 females (93.8%) and 3 males (6.2%), and 30 subjects (62.5%) had graduated from a nursing school/a junior college (3-year course), 16 (33.3%) had graduated from a 4-year college, while 2 completed a graduate school as their last academic record.

4.2. Efficacy of E-Learning Program

1) Changes in scores for cognitive domain and emotional domain/ scores for psychomotor domain

Changes in scores for the emotional and the cognitive domains over time and scores for the psychomotor domain are shown in Table 2. Scores for the cognitive and the emotional domains were significantly increased after program ($p < 0.01$).

2) Changes in test scores for each theme of e-learning

Changes in scores for each e-learning theme before and after completing the theme are shown in Table 3. For all the items, scores were significantly increased after e-learning ($p < 0.001$). Especially, scores for knowledge about disease characteristics rose higher than else scores.

4.3. Evaluation on Program Operability by Subjects

Evaluations on the program operability by subjects are shown in Table 4. As a learning method, 47 subjects (97.9%) rated e-learning as "good" or "relatively good". Regarding the easiness of taking a course, 40 subjects (85.5%) rated e-learning as "easy" or "relatively easy". There were 7 subjects (14.6%) who rated as "difficult to take" mainly because "it was difficult to write down contents and I wanted a text" or "I want to know my course record and past test results".

Table 2. Scores for the three domains.

	n	Mean ± standard deviation		p value
		Before e-learning	After e-learning	
Cognitive domain	48	27.5 ± 4.5	38.9 ± 5.0	<0.001***
Emotional domain	48	15.6 ± 2.5	16.7 ± 1.9	0.001**
Psychomotor domain	19		128.0 ± 17.9	
Wilcoxon signed-ranks test				

N = 48

**p < 0.01, ***p < 0.001.

Table 3. Scores for each e-learning theme.

Theme	Sub themes	n	Mean ± standard deviation		p value
			Before e-learning	After e-learning	
Disease management		44	7.2 ± 0.9	8.6 ± 1.0	<0.001***
Knowledge about disease characteristics	Metabolic syndrome	42	7.1 ± 0.8	9.1 ± 0.9	<0.001***
	Diabetes	41	6.9 ± 0.8	9.1 ± 0.8	<0.001***
	Diabetic nephropathy	40	6.8 ± 1.1	9.0 ± 0.7	<0.001***
	Ischemic heart disease	43	7.8 ± 0.9	9.3 ± 0.9	<0.001***
	Heart failure	44	6.7 ± 1.2	8.8 ± 1.2	<0.001***
	Stroke	43	7.8 ± 1.2	9.2 ± 0.8	<0.001***
	Self-management skill education	43	6.7 ± 1.5	8.8 ± 1.4	<0.001***
Care coordination		45	7.1 ± 1.0	9.0 ± 0.9	<0.001***
Communication		37	7.3 ± 1.0	9.1 ± 0.8	<0.001***
Mean score for all tests		32	7.1 ± 0.8	9.1 ± 0.6	<0.001***
t-test					

n = 48

***p < 0.001.

Regarding the duration of e-learning, 25 subjects (52.1%) rated as "appropriate" and other 22 (45.8%) rated as "too long". For the numbers of themes included in the e-learning, 14 subjects (29.2%) rated as "appropriate" and 34 (70.8%) rated as "too many". Those who replied that the duration of e-learning was "too long" or that the number of themes included in e-learning was "too many", also described "since the course was divided into short sessions, it was easy to focus on, but it is too hard to complete the all sessions within the term."

As voluntary comments, we had some distinctive answers, e.g. the learners said that they could repeatedly learn about disease management based on evidence, and they thought that they understood the processes of patient education and could utilize it in the practice.

5. DISCUSSION

5.1. Operability of Program

Among the subjects, 97.9% replied that e-learning was a good learning method, and 85.5% replied that it was easy to take, which indicates a high evaluation for the program quality. In addition, the completion rate of the program was 87.3% and all subjects who stopped the program did so because they were too busy. In this study, taking into account the operability, we set the time frame at 2 months. However, since some subjects commented in program evaluation that they felt it as a burden to complete the whole program within the limited time, the operability may be improved if we set a longer time frame.

5.2. Efficacy of Program

Telenursing is a concept that was defined by the International Council of Nurses in 1998, and there have been various reports abroad and in Japan on improvement of disease management and containment of healthcare expenses by telenursing in patients with a chronic disease [29] - [33] . ICN states that the skill of telenursing is a tool to perform nursing process including assessment, diagnosis, planning, exercise and evaluation for patients, and that the ability to perform nursing process as well as development of partnership influence the quality of telenursing. In addition, in order to perform disease management, not only evidence-based patient education but also the abilities to support practice and collaborate with other professionals are required [34] . We developed our learning program to train and evaluate nurses so that they could acquire knowledge and skills related to them.

The subjects of our program increased scores of e-learning for the emotional domain (interests in disease management and patient education) and the cognitive domain (subjective evaluation on acquiring knowledge about disease management and patient education). Also, for the psychomotor domain (patient education skills), we set items that could evaluate skills required for face-to-face and telenursing patient education. However, the evaluation for the psychomotor domain was made only after the program. Although there is the limitation of the study, we consider that since we achieved our intended results, the program was structured appropriately.

Table 4. Evaluation of the program by the subjects.

				N = 48 n (%)
Overall view of e-learning	Good	Relatively good	Not so good	Bad
	25 (52.1)	22 (45.8)	1 (2.1)	0 (0.0)
Easiness of e-learning	Easy	Relatively easy	Difficult	
	33 (68.8)	8 (16.7)	7 (14.5)	
Duration of e-learning	Too long	Appropriate	Too short	
	22 (45.8)	25 (52.1)	1 (2.1)	
Themes included in e-learning	Too many	Appropriate	Too few	
	34 (70.8)	14 (29.2)	0 (0.0)	

Knowls stated that in adult learning, it is important to utilize their own experience as abundant learning resources and to show how to apply them in the practice [35] . Also, there were voluntary comments by subjects such as "I could learn by comparing with my actual experience" and "I would like to make good use in everyday practice", and these comments indicate that they could learn deeply using they own experience and that they could obtain some suggestion for the actual practice. Therefore, we consider that repeated learning using experience and contents along with thinking process of experts based on the practice were effective.

Furthermore, it has been reported that in chronic disease management, education to improve patient's ability of self-management is most effective in improving patient's outcome [19] [36] , and it is essential for disease management. To understand how the intervention by nurses who enhanced their skills in patient education after our e-learning program could improve outcomes in patients with a chronic disease, it would be necessary to conduct further investigation.

6. LIMITATIONS OF THE STUDY AND FUTURE CHALLENGES

As the design of our study, we performed tests before and after only in an intervened group and did not set a control group, because the nurse administrators of the facilities desired that all nursing staff needed to learn the skills. Also, the evaluation term was short with 2 months, but it is considered that in order to evaluate skill acquisition, a longer term for before-after evaluation and comparison was necessary. However, the positive result from this study suggests that further investigation with a controlled comparison and a longer-term evaluation may demonstrate more valuable results.

7. CONCLUSION

We carried out a nurse training program using e-learning on disease management and patient education, and it improved subject's knowledge and interests. We consider that knowledge enhancement through e-learning and learning along with the process of patient education can contribute to improvement of patient education skills.

CITE THIS PAPER

KanaKazawa,MichikoMoriyama,MichiyoOka,SatsukiTakahashi,MadokaKawai,MasumiNakano, (2015) Efficacy and Usability of an E-Learning Program for Fostering Qualified Disease Management Nurses.*Health*,**07**,955-964. doi: 10.4236/health.2015.78113

REFERENCES

1. Barlow, J., Wright, C., Sheasby, J., Turner, A. and Hainsworth, J. (2002) Self-Management Approaches for People with Chronic Conditions: A Review. Patient Education and Counseling, 48, 177-187.

2. Lozano, R., Naghavi, M., Foreman, K., et al. (2012) Global and Regional Mortality from 235 Causes of Death for 20 Age Groups in 1990 and 2010: A Systematic Analysis for the Global Burden of Disease Study 2010. Lancet, 380, 2095-2128.

3. Suhrcke, M., Daragh, K.F. and McKee, M. (2008) Economic Aspects of Chronic Disease and Chronic Disease Management. In: Nolte, E. and Mckee, M.B., Eds., Caring for People with Chronic Conditions, Open University Press, England, 43-63.

4. Ministry of Health, Labour and Welfare (2014) Promotion of Data-Health Project. (In Japanese) http://tokuteikenshin-hokensidou.jp/interview/002/datahealth_mhlw.pdf

5. Fukuoka, Y., Hosomi, N., Hyakuta, T., Omori, T., Ito, Y., Uemura, J., Kimura, K., Matsumoto, M. and Moriyama, M., DMP Stroke Trial Investigators (2014) Baseline Feature of a Randomized Trial Assessing the Effects of Disease Management Programs for the Prevention of Recurrent Ischemic Stroke. Journal of Stroke and Cerebrovascular Diseases, 24, 610-617.

6. Kazawa, K. and Moriyama, M. (2012) Effects of the Educational Program for Pre-Dialysis Patients with Diabetic Nephropathy (Intervention Evaluation for 6 Months). The Journal of Japan Academy of Nephrology Nursing, 14, 92-100. (In Japanese)

7. Kazawa, K. and Moriyama, M. (2013) Effects of a Self-Management Skills-Acquisition Program on Pre-Dialysis Patients with Diabetic Nephropathy. Nephrology Nursing Journal, 40, 141-149.

8. Kazawa, K., Takeshita, Y., Yorioka, N. and Moriyama, M. (2014) Efficacy of a Disease Management Program Focused on Acquisition of Self-Management Skills in Pre-Dialysis Patients with Diabetic

Nephropathy: 24 Months Follow-Up. Journal of Nephrology, 28, 329-338.

9. Moriyama, M., Takeshita, Y., Haruta, Y., Hattori, N. and Ezenwaka, C.E. (2013) Effects of a 6-Month Nurse-Led Self-Management Program on Comprehensive Pulmonary Rehabilitation for Patients with COPD Receiving Home Oxygen Therapy. Rehabilitation Nursing, 40, 40-51.

10. Otsu, H. and Moriyama, M. (2012) Follow-Up Study for a Disease Management Program for Chronic Heart Failure 24 Months after Program Commencement. Japan Journal of Nursing Science, 9, 136-148.

11. Takami, C., Moriyama, M., Nakano, M., Kuroe, Y., Nin, K., Morikawa, H., Hasegawa, T. and Hayashi, S. (2008) Developmental Process of Disease Management Program of Type 2 Diabetes with a View to Acquiring Self-Management Skills: Effects of the Trial Implementation. Japan Journal of Nursing Science, 28, 59-68.

12. Fitzner, K., Sidorov, J., Fetterolf, D., Wennberg, D., Eisenberg, E., Cousins, M., Hoffman, J., Haughton, J., Charlton, W., Krause, D., Woolf, A., McDonough, K., Todd, W., Fox, K., Plocher, D., Juster, I., Stiefel, M., Villagra, V. and Duncan, I. (2004) Principles for Assessing Disease Management Outcomes. Disease Management, 7, 191-201.

13. Population Health Alliance (2014) PHM: Disease Management. http://www.populationhealthalliance.org/research/phmglossary/d.html

14. McCutcheon, K., Lohan, M., Traynor, M. and Martin, D. (2014) A Systematic Review Evaluating the Impact of Online or Blended Learning vs. Face-to-Face Learning of Clinical Skills in Undergraduate Nurse Education. Journal of Advanced Nursing, 71, 255-270.

15. Segal, G., Balik, C., Hovav, B., Mayer, A., Rozani, V., Damary, I., Golan-Hadari, D., Kalishek, S. and Khaikin, R. (2013) Online Nephlorogy Course Replacing a Face to Face Course in Nursing Schools' Bachelor's Program: A Prospective, Controlled Trial, in Four Israeli Nursing Schools. Nurse Education Today, 33, 1587-1591.

16. Hart, P., Eaton, L., Buckner, M., Morrow, B.N., Barrett, D.T., Fraser, D.D., Hooks, D. and Sharrer, R.L. (2008) Effectiveness of a Computer-Based Educational Program on Nurses' Knowledge, Attitude, and Skill Level Related to Evidence-Based Practice. Worldviews on Evidence-Based Nursing, 5, 75-84.

17. Moattari, M., Moosavinasab, E., Dabbaghmanesh, M.H. and ZarifSanaiey, N. (2014) Validating a Web-Based Diabetes Education Program in Continuing Nursing Education: Knowledge and Competency Change and User Perceptions on Usability and Quality. Journal of Diabetes & Metabolic Disorders, 13, 70.

18. Sheen, S.T.H., Chang, W.Y., Chen, H.L., Chao, H.L. and Tseng, C.P. (2008) E-Learning Education Program for Registered Nurses: The Experience of a Teaching Medical Center. Journal of Nursing Research, 16, 195-201.

19. Funnell, M.M., Brown, T.L., Childs, B.P., Haas, L.B., Hosey, L.B., Jensen, B., Marynuik, M., Peyrot, M., Piette, J.D., Reader, D., Siminerio, L.M., Weinger, K. and Weiss, M.A. (2007) National Standards for Diabetes Self-Management Education. Diabetes Care, 30, 1630-1637.

20. Bloom, B.S., Hastings, J.T. and Madaus, G.F. (1971) Handbook on Formative and Summative Evaluation of Student Learning. McGraw-Hill Book Company, New York.

21. Japanese Society of Nephrology (2013) Evidence-Based Practice Guideline for the Treatment of CKD. Tokyo Igakusha, Tokyo. (In Japanese)

22. The Japan Diabetes Society (2013) Evidence-Based Practice Guideline for the Treatment for Diabetes in Japan 2013. Nankodo, Tokyo. (In Japanese)

23. The Japanese Circulation Society (2010) Guidelines for Treatment of Chronic Heart Failure (JCS 2010). http://www.j-circ.or.jp/ guideline/ pdf/JCS2010_matsuzaki_h.pdf

24. The Japanese Circulation Society (2011) Guidelines for Secondary Prevention of Myocardial Infarction (JCS 2011). (In Japanese) http://www.j-circ.or.jp/guideline/pdf/JCS2011_ogawah_h.pdf

25. The Japan Stroke Society (2009) Japanese Guidelines for the Management of Stroke. Kyowa kikaku, Tokyo. (In Japanese)

26. Bandura, A. (1977) Self-Efficacy: Toward a Unifying Theory of Behavioral Change. Psychological Review, 84, 191-215.

27. Prochaska, J.O. and Velicer, W.F. (1997) The Transtheoretical Model of Health Behavior Change. American Journal of Health Promotion, 12, 38-48. http://dx.doi.org/10.4278/0890-1171-12.1.38

28. Morikawa, H., Nonomura, N. and Muranaka, Y. (2001) The Trends of Nursing with CAI and the State of Available Materials. Kangotenbo, 26, 68-75. (In Japanese)

29. Dunagan, W.C., Lit-tenberg, B., Ewald, G., Jones, C.A., Emery, V.B., Waterman, B.M., Silverman, D.C. and Rogers, J.G. (2005) Randomized Trial of a Nurse-Administered, Telephone-Based Disease Management Program for Patients with Heart Failure. Journal of Cardiac Failure, 11, 358-365. http://dx.doi.org/10.1016/j.cardfail.2004.12.004

30. Jensen, L., Leeman-Castillo, B., Coronel, S.M., Perry, D., Belz, C., Kapral, C. and Krantz, M.J. (2009) Impact of a Nurse Telephone Intervention among High-Cardiovascular-Risk, Health Fair Participants. Journal of Cardiovascular Nursing, 24, 447-453.

31. Kamei, T., Yamamoto, Y., Kajii, F., Nakayama, Y. and Kawakami, C. (2013) Systematic Review and Meta-Analysis of Studies Involving Telephone Monitoring-Based Telenursing for Patients with Chronic Obstructive Pulmonary Disease. Japan Journal of Nursing Sciences, 10, 180-192. http://dx.doi.org/10.1111/j.1742-7924.2012.00228.x

32. Kim, H.S. (2007) Impact of Web-Based Nurse's Education on Glycosylated Hemoglobin in Type 2 Diabetic Patients. Journal of Clinical Nursing, 16, 1361-1366.

33. Shearer, N.B.C., Cisar, N. and Greenberg, E.A. (2007) A Telephone-Delivered Empowerment Intervention with Patients Diagnosed with Heart Failure. Heart & Lung: The Journal of Acute and Critical Care, 36, 159-169.

34. Wagner, E.H. (2000) The Role of Patient Care Teams in Chronic Disease Management. British Medical Journal, 320, 569-572.

35. Knowls, M.S. (1988) The Modern Practice of Adult Education: From Pedagogy to Andragogy. Association Press, New York, 45-54.

36. Nakano, M., Moriyama, M. and Nishiyama, M. (2003) Structured Review of the Literature of Type 2 Diabetes Self-Management: Toward the Development of a Patients' Characteristics-Based Assessment Tool. Journal of Health Sciences, 3, 1-12. (In Japanese)

CHAPTER 7

A Knowledge-building Process in Interaction-based E-Learning

Hye-Jung Lee[1]

[1] *Institute for Education and Innovation, South Korea*

Abstract

This research articulates a knowledge-building process in interaction-based e-learning. For exploration of a knowledge-building process, an interaction-based e-learning program was developed and implemented at a college level course. Throughout the course, quantitative and qualitative data including students' perceived knowledge-building process from questionnaires, online messages, interview data, and participatory observation journal were collected and analyzed. As results, an observable action model and a conceptual model of the knowledge-building process were derived, which students and experts verified. Cognitive achievement factors and satisfaction factors were also considered in the knowledge building process model. Meaning and implication of each stage in the model were discussed.

Keywords

Learning process mechanism, CSCL, knowledge building, distance learning, interactive e-learning

1. INTRODUCTION

Education at a distance is becoming increasingly *interactive* with the ever sophisticated advances in web technology and therefore, interactive learning supported by the technology becomes a more significant field than ever, raising lots of critical issues in research and practice. Interaction-based e-learning may host many modes of communication, such as threaded discussion forums, chat, email, wiki-based boards, etc. Most of the research persists positive effects of interaction and present various strategies to improve the interaction for better learning [1, 2].

Although much of the research emphasizes the effectiveness of interaction and strategies to make the interaction more active and effective in an e-learning environment, why and how the interaction or strategies are effective has not yet been studied enough. There are some conceptual and theoretical articles on knowledge-building [3, 4]. However, there is a lack of theoretical research on a knowledge-building process based on empirical implementation. To find out more effective instructional strategies in interactive distance learning, we need to first understand how the knowledge-building process works in interaction-based e-learning[1] - .

Knowledge-building is said to be differentiated from *learning* [3, 5]. In [3], "Learning is an internal, unobservable process that results in changes of belief, attitude, or skill. Knowledge-building, by contrast, results in the creation or modification of public knowledge"(p. 1371). [3] also described that "knowledge-building environments enable ideas to get out into the world and onto a path of continual improvement in a form that allows them to be discussed, interconnected, revised, and superseded"(p. 1372). It focuses more on building knowledge-in-the-world as opposed to knowledge-in-the-head. To understand how the process mechanism is going and to find out better instructional strategy in interaction-based e-learning, the observable knowledge-building process rather than the internal learning process would be more useful. Also, the knowledge-building process should be more clearly disclosed in an interaction-based learning environment with active social communication, rather than resource-based instruction for basically individual learning.

The purpose of this chapter, therefore, is to articulate a knowledge-building process in interaction-based e-learning. This research is concerned with how individuals and groups build their knowledge and construct meaning in interaction-based e-learning. This research also considers learning outputs, such as cognitive achievement or satisfaction levels, for better comprehensive understanding of the knowledge-building process.

This research will focus on a process-oriented approach; such an approach is focused on '*where* it makes a difference' rather than '*whether* it makes a difference'. Many other studies that utilize a learning *product-oriented* approach—such as comparative studies of learning results with the application of certain strategies—showed various differences on their effectiveness. However, it is said to be no significant difference by meta-analysis of each research result [6]. This 'no significant difference phenomena' indicates that research needs to shift from finding differences to reasoning the cause of differences. This research focuses more on *process* than product, and presents a more meaningful contribution in the theory and practice of interactive e-learning.

2. THEORETICAL BACKGROUND

The learning process has been studied by learning psychologists in behaviorism, cognitivism, or constructivism; however, learning mainly occurs in our brain which is basically non-observable. Only the consequences of learning can be observed. So many studies have dealt with the learning consequences rather than the learning process. However, in many cases, those studies generally turn out to be statistically insignificant when one tries to be rigorous about the learning outcome [6]. Therefore, [5] suggests that we should use the term, knowledge-building rather than learning, especially in regard to collaborative and interactive learning. According to Stahl [4, 5], knowledge-building is more tangible, concrete, and descriptive than learning. This term, knowledge-building, seems to include the whole process of external activity influencing on learning as well as internal learning itself within the brain. With care and practice, the knowledge-building process can be observed directly and empirically, because it accounts for externally observable activities and artifacts as experiential evidence. Therefore, we will use the term knowledge-building instead of learning to specify the observable and empirical approach to this research.

One who tried to disclose the knowledge-building process in interactive e-learning, such as Computer-supported Collaborative Learning (CSCL), was Stahl [4]. Stahl presented a diagram of a knowledge-building process in CSCL from theoretical discussion. His diagram consisted of two circles: one of personal understanding and the other of social knowledge-building (see Figure 1). Stahl [4] described the diagram as *"The convention in the diagram is that arrows represent transformative processes and that rectangles represent the products of these processes: forms of knowledge. To take this limited representation too seriously would be to reify a complex and fluid development—to put it into boxes and to assume that it always follows the same path. In particular, the diagram gives the impression of a sequential process whereas the relations among the elements can take infinitely varied and complex forms. Indeed the identification of the particular set of elements is arbitrary and incomplete. Perhaps despite such limitations and potential distortions the diagram can provide a starting point for discussing a cognitive theory of computer support for knowledge-building. It remains to be seen if such a phase model provides the most useful representation (In [4], pp.71)."*

He explicitly considered the relationship of processes associated with individual minds to those processes considered to be socio-cultural. The significance of Stahl's model is that he indicates the importance of social learning, which is considered to be essentially different from individual self-learning. He suggests that knowledge would be shared and constructed by social interaction in a CSCL environment. He is taking a social constructivist's perspective in which his work impresses upon a sequential process to knowledge-building and provides a starting point for discussing cognitive theory of CSCL as indicated in his research [4]. However, his model was derived from theoretical discussion and it wasn't verified by empirical evidence. As he mentioned in several papers [4, 5, 7], the research community should elaborate upon the knowledge-building process model by utilizing empirical research.

Besides, many studies on modeling the learning process have been reported [8]. Most of them, however, use a face-to-face learning environment or do not utilize empirical evidence. Moreover, they present linear learning procedures and do not consider other factors such as influential relationship between process and product. Therefore, a study on a comprehensive knowledge-building process that considers the learning output variables, such as cognitive achievement and satisfaction levels, as a form of empirical evidence is needed.

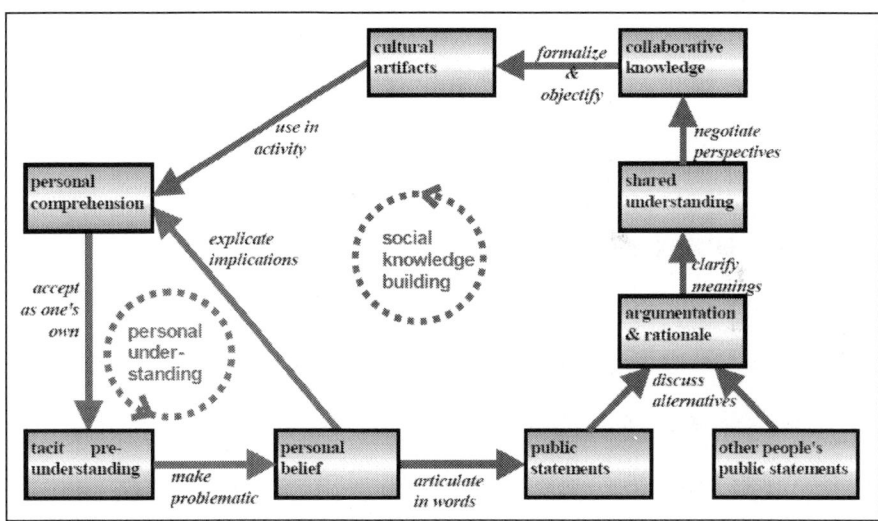

Figure 1. A diagram of the knowledge-building processes [4].

3. METHOD

In order to articulate a knowledge-building process, an interaction-based e-learning course at college level was developed. Four experts in instructional design practice verified the course program throughout the development process. The subject of the program was 'General Understanding of Distance Education'. Fifty-six juniors at K Cyber University in Korea, from ages 20 to

50, were required to partake in team interactions assigned in the class for a period of eight weeks. Throughout the course, the students' perceived knowledge-building process was collected and analyzed from questionnaires. Cognitive achievement tests, satisfaction queries, online messages, interviews, and participatory observation data were also collected and analyzed. Various methods of statistical analysis including correlation analyses, contents analyses, t-tests, frequency analyses, etc. were applied to the data as well. For students' perceived knowledge-building process, each student was required to describe his/her own knowledge-building process four times during the course. Four times investigations helped how students' perception changed throughout the course progress. Average return rate of the each questionnaire was 84% out of all 56 students. Question items were as follows:

- How many hours per week do you spend for this course?

- Which learning step do you spend the most time on? Write your answers in order.

- What influences your examination the most?

- While working on an assignment, what do you rely on the most?

- What is the most critical satisfactory factor in this course?

- What was dissatisfactory in the course?

- How often do you access to the web class?

- Describe your knowledge-building process in order. Write every single visit and activity on the web class site during your stay as detailed as possible.

For evaluation reliability, three evaluators graded 10% of the students' answer sheets and their responses were correlated (Pearson $r=0.84$, $p<0.01$). In terms of satisfaction level, a satisfaction measurement tool that was developed and validated by [9] was used after modification (reliability alpha=0.93).

After the course, 10 students were interviewed by telephone for 1–2 hours to verify all the quantitative data and to provide more detailed information regarding the factors of the knowledge-building process. The interview first started with the questions similar to the questionnaires in a flexible and unstructured manner and proceeded naturally to verify the knowledge-building models (an observable action model and a conceptual model reflecting the action model and other previous literature). All interview data were recorded and analyzed afterwards.

In the end, the conceptual model was verified by five experts (Ph.Ds. in Education) and ten students in the class. Respondents used a 5-point Likert scale (5 = fully verified, 1 = not verified), which was developed based on previous literature [1, 10] to rate validity, explicability, usability, generality, and comprehensibility of the conceptual model. Average rate of experts was 4.30 and average rate of students was 4.11 out of 5.00.

This study followed Rubinstein's [11] proposed procedure for modeling the knowledge-building process. According to Rubinstein, the modeling procedure is to achieve a simple high level of abstraction. So the procedure needs to be iterative until we get an abstractive pattern. In considerations of Rubinstein's perspective, the procedure in this research was as follows; Development of a treatment instructional program → Implementation of the program and collecting data → Coding questionnaire data → Deriving a rough pattern of the knowledge-building process → Analysis of learning output variables → Correlation of learning output variables and each stage of the knowledge-building process → Verification of the knowledge-building process action model by interview with learners → Conceptualization of knowledge-building process and its actual visualization → Verification of the conceptualized knowledge-building process model by experts and learners → Production of the verified conceptual diagram of the knowledge-building process in interaction-based e-learning. The more detailed procedure is shown in Table 1.

Table 1. Research procedure of this study.

The modeling procedure of the Learning Process in this study	Product
Develop a treatment instructional program	Interaction-based e-learning program
↓↓	↓↓
Implement the program and collect data	4 questionnaires and online message analysis
↓↓	↓↓
Make first coding of four questionnaires	17 steps of the knowledge-building procedure
↓↓	↓↓
Make second coding of four questionnaires	10 main stages and some sub-steps of the knowledge-building procedure
↓↓	↓↓
Derive a rough pattern of the knowledge-building procedure	Visualization of the knowledge-building procedure
↓↓	↓↓

The modeling procedure of the Learning Process in this study	Product
Analyze learning output variables	Student achievement, satisfaction
↓↓	↓↓
Correlate learning output variables and each stage of the knowledge-building process	First visualization of an action model of the knowledge-building process in consideration of learning output variables
↓↓	↓↓
Implement interview with learners	Interview recording data with 10 learners
↓↓	↓↓
Analyze interview results to verify the model	Decoding the recording, →→ contents analysis, →→ first categorization, →→ coding and second categorization
↓↓	↓↓
Derive a verified action model of	Visualization the action model of the knowledge-building

The modeling procedure of the Learning Process in this study	Product
the knowledge-building process	process
↓↓	↓↓
Conceptualize the knowledge-building action process	Visualization of the conceptualized knowledge-building process
↓↓	↓↓
Validate the conceptual model of a knowledge-building process	Validation the conceptual model of a knowledge-building process by 5 experts and 10 students
↓↓	↓↓
Produce the verified conceptual model of the knowledge-building process	The verified conceptual model of a knowledge-building process in interaction-based e-learning

4. RESULTS AND DISCUSSION

4.1. An Observable Action Model

To explore a knowledge-building process in interaction-based e-learning, we coded 56 students' perceived learning procedures in questionnaires and derived an average pattern of the students' knowledge-building process. The students' perceived learning procedure after first coding was composed of 17 stages as follows:

1. reading notices & information of the department

2. reading notice of the course

3. reading the Q/A board

4. posting messages on the free board or the Q/A board

5. reading messages on the discussion board

6. studying web-based material

7. editing and printing web-text

8. doing assignments

9. searching other materials for reference

10. reviewing peers' posting and teacher's feedback on it

11. posting questions or replies on the discussion board

12. offline interaction (telephone or offline meetings)

13. assignment submissions

14. checking the teacher's feedback

15. reflection

16. resubmitting assignments after revision

17. checking my individual learning pace in the LMS (learning management system)

Among these items, activities receiving less than 10% frequency of use were removed after first coding, and the learning procedure was re-coded iteratively until an average main pattern of the process was found. Thus, #1, #12, and #17 items were removed, and other items were relocated to the basic default procedure of #6, #8, #13 cycle; studying web-based material, doing assignments, assignment submission. Finally, ten main stages and some sub-steps were induced. For main stages: #6, #8, #13 basic cycle (studying web-based material, doing assignments, assignment submission); #14, #15, #16 activity (checking the teacher's feedback, reflection, resubmission after revision); #4, #5, #10, #11 activity (posting messages on the free board, Q/A board, discussion board, reviewing peers' posting and teacher's feedback on it, posing questions and replies on the discussion board). Sub-steps that students do sometimes but are not that critical according to the frequency are #2, #3, #7, and #9 (reading notice, Q/A board, editing and printing web-text, searching other materials for reference). All stages were analyzed and correlated with learning output variables such as cognitive achievement testing or satisfaction queries. In addition, all messages on each web board were analyzed and categorized by characteristics of message content, SDU (Social Discussion Unit), PDU (Procedural Discussion Unit), and CDU (Contents Discussion Unit), following the classification of [12]. An observable action model of knowledge-building process, in which all stages were rearranged with a logical sequence, was finally derived as shown in Figure 2.

In Figure 2, subscript 1) represents a cognitive achievement factor and subscript 2) indicates a satisfaction factor. Subscript 3) shows features of messages, such as SDU (Social Discussion Unit), PDU (Procedural Discussion Unit), and CDU (Contents Discussion Unit), categorized by [12]. Subscript 4) represents the form of interaction, such as S-C (Student-Contents), S-T (Student-Teacher), S-S (Student-Student), categorized by [13].

In the student-contents interaction (S-C) circle, students come in contact with the web material and are then involved in the process of doing the assignments. While students process their assignments, they interact with other students. This kind of action leads to the student-student interaction (S-S) circle. Meanwhile, when students get feedback from their instructor after submitting their assignments, they check and reflect the teacher's feedback. These steps are for production of assignments. This kind of action is categorized as student-teacher interaction (S-T). Throughout this entire process, students read notices and information on the bulletin board concurrently.

In the student-student interaction (S-S) circle, students referred to peers' finished assignments, read messages on the discussion board, and post messages that are social, procedural, and academic in characteristic. Students were able to see other classmates' finished assignments because all students were supposed to post their assignments on an open discussion board for this research.

In the student-teacher interaction (S-T) circle, after submitting their assignments, students receive and review the teacher's feedback on their work. After reflection, some students revised their assignments and resubmitted them.

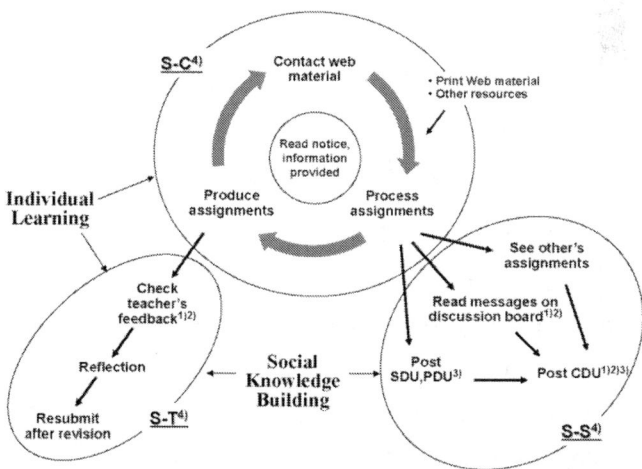

Figure 2. An observable action model of a knowledge-building process in interaction-based e-learning: 1) cognitive achievement factor; 2) satisfaction factor; 3) sdu(social discussion unit), pdu(procedural discussion unit), cdu(contents discussion unit); 4) interaction types (students-contents(s-c), students-students(s-s), students-teacher(s-t).

Regarding posting messages, the numbers of postings of each student were analyzed with achievement score by correlation analysis. Only CDU was significantly correlated to achievement score by r=0.455(p<0.05, N=52), and to satisfaction score by r=0.407(p<0.01, N=52). As expected, posting SDUs or PDUs did not show significant correlation with cognitive achievement.

Satisfaction level result measured by a modified satisfaction scale of [9] 's was correlated to each stage of the action model in Figure 2, and found significant correlations only with "check the teacher's feedback", "read the messages on the discussion board", and "post CDU"(p<0.05). With respect to the cognitive achievement factor, the students who checked the teacher's feedback showed significantly higher scores than the students who didn't check the teacher's feedback (Table 2). Table 3 shows that students reading messages on the board had higher scores in the final examination than students who did not read the messages.

Table 2. T-test result: whether checking the teacher's feedback is a critical achievement factor.

Group	N	Mean	St. Dev.	df	T
Students who checked the teacher's feedback	19	66.63	21.10	24.07	2.43
Students who did not check the teacher's feedback	33	79.33	11.31		

p< 0.05

Table 3. T-test result: whether checking the teacher's feedback is a critical achievement factor.

Group	N	Mean	St. Dev.	df	T
Students who read the messages on the board	11	65.82	23.09	50	2.06
Students who did not read the messages on the board	41	79.07	13.80		

'p< 0.05

Reading messages on the discussion board, posting CDUs, and reviewing the teacher's feedback are figured as satisfaction factors by frequency analysis and correlation analysis ($p<0.05$). Unexpectedly, the student achievement factors are the same as satisfaction factors in this case, but this could not always happen in other cases. It needs further investigation to differentiate the influence of achievement and satisfaction.

4.2. A Conceptual Model

The observable action model is abstracted into a conceptual diagram (Figure 3) in consideration of previous research [4, 14]. In Figure 3, the bold solid arrows show a major knowledge-building process and the fine solid arrows show a back process or a minor process that did not occur all the time. The conceptual model of the knowledge-building process in interaction-based e-learning constitutes two phases: a minor individual learning phase and a major social knowledge-building phase. Even though social learning is a major part of the knowledge-building process in interaction-based e-learning, individual learning occurs almost concurrently or alternately with social learning. Although the instructional program in this research was designed mainly for interactive learning, students experienced self-learning with brief material provided in the class to learn basic information for discussion preparation. So an individual learning cycle must be shown with the social knowledge-building cycle concurrently. Explanations of each stage are described below.

4.2.1. Initiation

In the individual learning phase, the process begins with initiation. Initiation includes access and exploration of the program sites such as reading notices or announcements, clicking menu options, etc. But this activity is not a main learning process; rather it is a pre-activity before the learning process. Once students get used to navigating their way around the site, they usually skip this stage and go to the main learning process directly. So the initiation phase is located outside of the learning cycle.

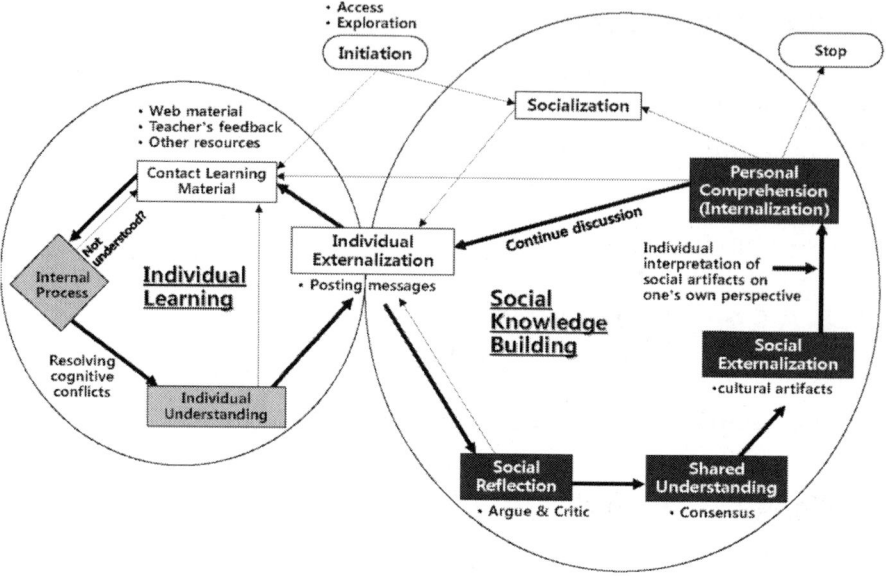

Figure 3. A conceptual model of a knowledge-building process in interaction-based e-learning ([   ]: from empirical evidence, [   ]: from previous literature, [   ]: from previous literature and empirical evidence; : major knowledge-building process, : minor knowledge-building process).

4.2.2. Contact With The Learning Material

After initiation, learners come in contact with the learning material that includes the web-based instruction program, ongoing teacher's feedback, or other online/offline resources. The reason why we differentiated the stage of contact with the learning material from other stages such as internal process and understanding is because even though students come in contact with the web lecture, it doesn't mean that they are really engaged in learning. If students enter the web lecture and click each page of the web material, it is possible that they are just clicking through the program, which is far from genuine learning. So this stage of coming in contact with the learning material should be differentiated for articulation.

4.2.3. Internal Process

When learners digest the learning material, an internal process must take effect in their brain, which is not observable but explained by many learning theories. This is for intra-personal communication represented by thought. What's going on in this internal process will not be discussed here because it is beyond this study's scope, but it is obviously appropriate to put this step as one of the stages in the learning process model here. If students understand the contents well enough after the internal process, they arrive at an individual understanding. Otherwise, they return to the learning material and repeat this cycle until they understand the material.

4.2.4. Individual Understanding

If a learner's cognitive conflicts are resolved through this internal process, students arrive at an individual understanding. Similar to Stahl's model [4], this is distinguished from personal comprehension by internalization. Although we guess that we understand something, we often find that we are unable to explain what we have learned immediately. That is because the knowledge is not fully internalized yet, though it may be slightly understood. Individual understanding, therefore, could be considered a lower level of comprehension. That is, knowledge is not internalized to a learner yet in the stage of individual understanding.

4.2.5. Individual Externalization

In the e-learning course in this study, students post what they learned from the material after individual-understanding; this action is conceptualized as individual externalization in this research. Students also post messages following socialization or internalization on the knowledge-building cycle. That is, students express what they learned from social learning or individual learning. So this stage shows two facets: one is a summary of the individual learning phase and the other is the first step to the knowledge-building phase, which also follows the socialization or internalization process. The stage of individual externalization seems to be similar to making personal belief elicit to public statements in Stahl's model [4].

4.2.6. Socialization

Students participate in discussions by posting messages of what they learned through individual learning or just by socialization. In the knowledge-building phase, students begin to take part in the discussions by posting social messages (SDUs) or asking about procedure (PDUs); this non-academic activity is for their social affinity and rapport. This is conceptualized as socialization in this study. This does not always happen. Once students are socialized enough (SDU), they usually skip this stage and go straight to posting CDUs (individual externalization) after final personal comprehension of one thing. In this study, only 16% of the messages were SDUs. They are shown as fine solid arrows rather than bold arrows in order to represent a minor process. After socialization, students post content-related academic messages (CDUs) on the discussion board, which is expressed as individual externalization, as mentioned above.

4.2.7. Social Reflection

When several individuals' messages are posted on the web board, students argue and criticize others' opinions. That is called social reflection in this study. Stahl [4] presents this stage as critic and argument of other people's public statements and discussion of alternatives. Social reflection is a corresponding concept to individual reflection; while one is from inter-personal interaction and the other is from intra-personal interaction, both are basically similar activities in regard to learning in a precise and concrete manner.

4.2.8. Shared Understanding

Through social reflection, students obtain consensus on a topic to arrive at shared understanding. Shared understanding is distinguished from individual understanding. This stage implicates that meaning is constructed by social practice as [4]. Social constructivists assert that meaning is constructed by social interaction until people share a common understanding. Shared understanding is from interpersonal interaction, whereas individual understanding is from intra-personal interaction.

4.2.9. Social Externalization

When one of the team members summarizes his/her cultural artifacts—product of discussion, summary of consensus like Stahl [4] mentioned—this activity is conceptualized as social externalization that is differentiated from individual externalization. While individual externalization consists of activities such as note-taking or summarizing of what students understand individually, social externalization consists of external expression of socially constructed and shared understanding. Usually one of the team members posts his/her summary or conclusion after discussion, while other team members watch and apply corrections if there is something incorrect or missing.

4.2.10. Internalization

Finally, students internalize knowledge into their personal comprehension schema. How knowledge is internalized into a personal comprehension schema after social externalization is one of the critical issues. In [7] on Meaning and Interpretation, he indicates that meanings in computer-supported collaborative learning are necessarily shared and must be interpreted by individuals. That is, learners interpret social artifacts, which are constructed by social interaction, on each individual's own perspective to reach personal comprehension. This is the only stage of intra-personal and non-observable stage in social knowledge-building, whereas other stages are mostly inter-personal and observable.

Another piece of empirical evidence of this process is that students who only read messages were found not to be inferior to students who write and post messages in terms of cognitive achievement (p<0.05). In this research, reading messages as well as writing and posting messages were found to be a critical achievement factor and a satisfaction factor. This finding implicates that one can get meaningful learning though he/she doesn't partake in social externalization after obtaining shared understanding; if one is not a team representative who is summarizing their discussion, he/she is hardly able to get an opportunity to externalize what they share from social reflection and just to watch and read other's externalized messages. In spite of not partaking in social discussion, these students showed high cognitive achievement just like those who posted social externalization messages. This means that there is some process for those who don't undertake observable externalization. It may be explained that people internalize their shared understanding by interpreting of social artifacts with each individual's own perspective to reach personal comprehension.

Regarding satisfaction, there was no significant difference between students who actively participated in social discussion by writing and posting messages and students who only read messages on the board ($p<0.05$). This implicates that students who only read messages also get meaningful learning and satisfaction through the dynamic interaction in the web class and there must be a certain stage to go to internalization. One study [15] gives a significant implication in this context. In [15], perception of overall interaction was a critical predictor of satisfaction. They suggested that overall dynamics in interaction might have a stronger impact on learners' satisfaction than strict personal participation. That is, vicarious interaction within the class as a whole than overt engagement of each participant may result in greater learner satisfaction. Therefore, reading only in a certain period of knowledge-building process could be considered as a meaningful learning activity.

5. CONCLUSION

This research was conducted to explore and articulate the knowledge-building process in interaction-based e-learning. For exploration of a knowledge-building process, an interaction-based e-learning program was developed and implemented at a college level course. Throughout the course, quantitative and qualitative data including the students' perceived knowledge-building process from questionnaires, online messages, interview data, and participatory observation journal were collected and analyzed. As results, an observable action model and a conceptual model of the knowledge-building process were derived, which students and experts verified. Cognitive achievement factors and satisfaction factors were also considered in the knowledge-building process model. Meaning and implication of each stage in the model were discussed.

The significance of this research is as follows: first of all, this study provides a conceptual framework for understanding a knowledge-building process in interaction-based e-learning such as online discussion learning. The model presented here was better articulated and elaborated based on empirical evidences, indicating social knowledge-building cycle is critically important in interaction-based learning. Historically learning has believed to occur within one's brain from an ultimately individual process. The results in this research provided a conceptual framework saying that there are two cycles and the social learning process as well as individual learning process is critical in interaction-based e-learning. Personal cognition and social activity might not be able to be separated artificially like the model in this research. Stahl [4] indicates by citing Hegel that it is the nature of a relationship of mutually constituting subjects: neither can exist without the other. But this kind of sequential visualization provides a more useful and clearer understanding of the knowledge-building process in interaction-based e-learning.

Second, this article provides a beginning to explain a mechanism of certain effectiveness of observable phenomena. For example, the report in [15] that vicarious interaction was a significant predictor of learning output could be correlated with the result that reading—as well as writing—messages on the discussion board was an achievement factor and a satisfaction factor. It is possible to explain that people learn sometimes by only reading even in interactive learning because people interpret social artifacts from socially shared understanding with an individual's own perspective to get to personal comprehension.

Third, several instructional design strategies such as externalization or group dynamics can be recommended. An interesting implication of this research is that if a student does not externalize either individually or socially, he/she cannot internalize the knowledge properly. This implies that instructors need to design *externalization* of a student's understanding as a requirement such as a 'Reflection paper' or 'Today's learning note', etc. Students who did not post messages actively during the class discussion could be required to wrap up the fierce discussions in order to experience group dynamics as well as externalization.

Besides, the knowledge-building process research in a resource-based self-learning environment or their comparative study can be proposed for further study. Research of more cases in various learning contexts or considering more various learners' characteristics will also contribute to elaborate and generalize the model presented in this research and will enrich understanding of the theory and practice of interactive e-learning.

NOTES

[1] - Since there is a kind of e-learning that is for self-learning based on e-contents with little to no interpersonal interaction, this research specifies interaction-based e-learning to focus more on highly interactive e-learning. Computer and network technology have contributed into two ways in teaching and learning; one is for developing effective contents, and the other is for enhancing interpersonal interaction. Contents-based self-learning, pretty popular in Korea, is based on one-way e-contents mostly in the form of VOD (video-on-demand) or WBI (web-based instruction). To eliminate the contents factor, the course developed in this research was interaction-based, with the minimum pre-developed contents.

REFERENCES

1. Lee H.-J., Rha I. Influence of 'Structure' and 'Interaction' on Student Achievement and Satisfaction in Web-Based Distance Learning. Educational Technology and Society. 2009;12(4):372-382.
2. Kanuka H., Rourke L., Laflamme E. The influence of instructional methods on the quality of online discussion. British Journal of Educational Technology. 2007;38(2):260-271.
3. Bereite, C., Scardamalia M. Technology and literacies: From print literacy to dialogic literacy. In N. Bascia, A. Cumming, A. Datnow, K.

Leithwood, & D. Livingstone (Eds.), International Handbook of Educational Policy. Dordrecht, Netherlands: Springer. 2005; p.749-761).

4. Stahl G. A Model of collaborative knowledge-building. In B. Fishman & S. O'Connor-Divelbiss (Eds.), Fourth International Conference of the Learning Sciences Malwah, NJ; Erlbaum. 2000; p. 70-77.

5. Stahl G. Contributions to a theoretical framework for CSCL. In: Proceedings of the Conference on Computer Support for Collaborative Learning: Foundations for a CSCL Community. International Society of the Learning Sciences. 2002. Available from http://gerrystahl. net/publications/conferences/2002/cscl2002/cscl2002.pdf [Accessed: 2015-07-11].

6. Russell T. The no significant difference phenomenon: A comparative research annotated bibliography on technology for distance education: As reported in 355 research reports, summaries and papers. North Carolina State University, 1999. Available from http://www. nosignificantdifference.org/ [Accessed: 2015-07-11].

7. Stahl G. Meaning and interpretation in collaboration. In: Proceedings of the International Conference on Computer Supported Collaborative Learning. 2003. Bergen, Norway.

8. Brown R.E. The process of community-building in distance learning classes. Journal of Asynchronous learning networks. 2001;5(2):18-35.

9. Kim J., & Ryu, H. Evaluation of teaching and learning system. Korea National Open University Press: Seoul; 2000.

10. Rha I., Hong S. A study on exploring at the process model of online learning community development. Journal of Educational Technology. 2004;19(3):101-122.

11. Rubinstein M. Patterns of Problem Solving. Englewood Cliffs. NJ: Prentice-hall, Inc. 1975.

12. Oren A., Mioduser D., Nachmias R. The development of social climate in virtual learning discussion groups. International Review of Research in Open and Distance Learning. 2002;3(1). Available from http://www.irrodl.org/index.php/irrodl/article/view/80/154 [Accessed: 2015-07-11].

13. Moore M.G. Three types of interaction. American Journal of Distance Education, 1989;3(2):1-6.

14. Lee H.-J. New perspectives of theoretical research in web-based distance education: Beyond Moore's Concepts. Korean Journal of Educational Research. 2004;42(1):137-168.

15. Fulford C.P., Zhang, S. Perceptions of interaction: The critical predictor in distance education. The American Journal of Distance Education. 1993; 7(3):8-21.

CHAPTER 8

3D Interactions between Virtual Worlds and Real Life in an E-Learning Community

Ulrike Lucke and Raphael Zender

Department of Computer Science, Chair for Complex Multimedia Application Systems, University of Potsdam, August-Bebel-Straße 89, 14482 Potsdam, Germany

ABSTRACT

Virtual worlds became an appealing and fascinating component of today's internet. In particular, the number of educational providers that see a potential for E-Learning in such new platforms increases. Unfortunately, most of the environments and processes implemented up to now do not exceed a virtual modelling of real-world scenarios. In particular, this paper shows that Second Life can be more than just another learning platform. A flexible and bidirectional link between the reality and the virtual world enables synchronous and seamless interaction between users and devices across both worlds. The primary advantages of this interconnection are a spatial extension of face-to-face and online learning scenarios and a closer relationship between virtual learners and the real world.

1. INTRODUCTION

Interactivity is closely related to aspects of networking and interdisciplinary development, bringing together researchers from engineering, computer science, media art and design, and social sciences. Here, the importance of computer science needs to be emphasized along with the growing immersion of digital systems in our daily life: dealing with computers can be seen as a new cultural technique besides reading, writing, and calculating [1, 2]. There is a strong mutual penetration of the digital and physical world, leading to phenomena like virtual reality (computers mirroring the real world) or augmented reality (real-world objects enriched with digital information). For several years, virtual 3D worlds gained significant public attention. The most recent and most famous of these worlds is Second Life, but several text- or graphic-based environments in the web existed before. As first euphoria and commercial initiatives calm down [3], the scientific interest in media and information theory is raising (e.g., for educational applications [4, 5] and for the problem of multiple identities [6]).

From a cultural or psychological perspective, virtual 3D worlds allow to study human behaviour in a decoupled, reversible way—like mirroring the reality, including other people's intimate thoughts, by interacting through a the 3D interface [7]. In this way, there is a significant change in that virtual items with an artificial, digital environment become more and more reality [8]. This opens a new perspective to the pervasiveness of human-computer interfaces: It does not matter where, how, and what type of interface the user is interacting with; his intuitive movement (input like turning around, entering an area, or touching the screen) is immediately followed by a reaction (output like showing a text or an image, playing an audio or video stream). The media architecture connects people, space, and data by interleaving physical and virtual reality creating an extended sphere of (inter)action. This relates to the theory of cognition where receiving, processing, and transmission of information are understood as a sensuous, somatic experience [9]. From this point of view, the tangibility of an object is not restricted to physical environments but extends into virtual worlds—provided that these offer users the possibilities to (inter)act like in real life.

This contrasts strongly to the common vision of tangible user interfaces [10] as graspable instantiation of graphical user interfaces. But, at a closer look it is the implementation of the same concept: to transparently couple bits and atoms in order to let the users interact with a computer in the same way they do with the real world.

Considering this, a systematic combination of real life and virtual interaction is promising a huge benefit for electronic learning, in terms of (not only virtually) tangible E-learning interfaces that enrich the experiences of learners—and probably also those of teachers. By a felt-as-somatic interaction with the learning environment the cognitive capabilities of students can be exhausted to a much larger extent than in traditional classroom settings, where learners are typically acting in a much more passive and less individual way. In the following, this paper demonstrates how the tangibility of real-life objects can be closely interweaved with elements in a virtual 3D world. The goal of our work was to systematically interconnect classroom and virtual learning in order to provide a higher level of individuality and flexibility to the user—not only in terms of a "3D remote control," but as a generalized architecture for flexible, bidirectional interchange of media between different environments (like classroom, media lab, learning platform, and virtual world). The many technical facets of our architectural framework are outside the focus of this paper, like service mash-ups [11], context awareness [12], and streaming media [13]. We will rather concentrate on integrating real-life settings and the virtual world Second Life. Thus, the main contribution of this paper is to explain the benefits of this approach for education. We are going to describe how the virtual and real-life environment fit into the general concept, which possibilities for interaction they offer to the users, how they can be interconnected, how we designed the 3D user interface, and finally which use cases become possible. Thus, the paper combines several perspectives from computer science, social sciences, and media design in an interdisciplinary approach.

2. RELATED WORK

A widely accepted model for interconnecting different teaching/learning settings is blended learning [14], where several phases of face-to-face and online learning are arranged in a predefined order. However, teachers and students are bound to certain phases and platforms/tools at every moment of a course.

We believe that the direct interconnection of different educational settings allows for a seamless combination of synchronous scenarios during the lecture (with interaction between teachers and learners) and asynchronous scenarios before and after (individual or collaborative preparation and wrap-up)—regardless of the used platform. This enhances the learning comfort, increases the scope and quality of a lecture, and advances mobility and equality of opportunities for learners and lecturers. Current developments in this area can be divided into two groups: point-to-point connections and systematic redesigns.

There exist a number of dedicated point-to-point connections between different platforms and tools.(i)Based on the practical need to simplify and accelerate the processes to deploy teaching and learning material, there have been some developments to automatically integrate lecture recordings into learning management systems [15, 16]. These solutions strongly rely on the used tools and techniques (i.e., recording software and learning platform).(ii)To provide another example, there are some mash-ups between virtual worlds and other platforms, like for 3D visualization of large data sets [17] or for establishing links from the 3D world to traditional content on a learning platform [18]. Also, there is no general approach behind these point-to-point connections.

All these solutions suffer from limited extensibility and complex maintenance due to their dependence from tools and technologies.

Systematic integrations of different platforms are rare, no matter if designed for education or for other purposes.(i)An abstract specification of educational presentation systems helps to identify related components and methods, for example, using an object-oriented model [19].(ii)Basic features of a learning platform can be identified and invoked from various external sources [20] instead of implementing isolated solutions in every learning platform or web portal, again.(iii)A generic middleware [21] can help to interconnect online games that are based on multiple platforms.

Usually, these solutions follow an approach of fundamental platform decomposition for a later flexible recombination of modules. Here, object-, service-, or peer-to-peer-based architectures come into play. However, this is hard to realize with existing tools and infrastructures.

Considering state-of-the-art design principles and sustainability of developments, a systematic integration is desirable. Services have proven to be a valid mechanism for enhancement of existing platforms [22], if applied in a coarse-grained manner.

3. BIDIRECTIONAL INTERCONNECTION OF FACE-TO-FACE AND VIRTUAL LEARNING

3.1. A Service-Oriented Architecture for Cross-Platform Media Distribution

Distributed application scenarios consist of a high number of tools, platforms, and infrastructures. Especially, network-based environments are characterized by a high degree of heterogeneity and dynamics, which requires a systematic approach for conception and implementation of a well-suited architectural model. Otherwise, performance, scalability, and long-term sustainability cannot be guaranteed. Recent developments often show an unstructured aggregation of dedicated point-to-point connection between specific systems—though the theory of distributed systems offers a pool of general models for different requirements and conditions of the application scenario [23]. To mention just a few popular architectures: client/server, publisher/subscriber, peer-to-peer, or broker architecture. The pros and cons of these models affect issues like required knowledge on communication partners, existence of bottle necks, or completeness and timeliness of responses.

Selection of an appropriate model comes along with a systematic analysis of the application scenario, usually with the help of formal models for actors, use cases, components, and processes. Depending on the nature of the application, this can be achieved, for instance, with graphical modelling techniques [24] as well as using algebraic structures [25]. Besides computer scientists, this involves several representatives of the application scenario in system development, and thus reduces the risk of technically driven aberrations. Especially, the broker model—as the concept behind the service oriented architecture (SOA)—requires substantial process modelling by domain experts, complementary change management from a social and organizational point of view, and a so-called SOA governance by the upper management level [26].

We chose the broker model for systematic interconnection of different interaction spaces mainly because of its high degree of heterogeneity, agility, scalability, and transparency. Therewith, we dynamically redirect interactions between different locations (represented as media and control services) without predefined knowledge on any site. Figure 1 shows this scenario using the example of telelecturing. Communication between involved parties basically takes place in three steps. (1) A provider registers its (lecturing) services along with their characteristic at the broker. (2) A consumer (prospective participant) asks the broker for a desired service (lecture). The broker returns information on location, syntax, and terms of use of the currently best-suited service. (3) The consumer connects to the provider and invokes its service.

This architecture unifies the interaction between different types of providers, consumers, and services, as all steps described above are based on abstract service descriptions and platform-independent communication. Consumer and provider are just required to implement a minimal service interface. Thus, existing infrastructures and tools do not need to be redesigned.

In an educational application scenario, there are some additional points that allow or even require the use of SOA. First of all, there are a number of established network addresses that are known to all clients and servers (like learning platforms); they can be used as brokers. Moreover, the content of a lecture typically is not security relevant, which simplifies the implementation and practical use of a prototype. (Nevertheless, a cross-institutional scenario requires basic services for authentication and accounting.) Another aspect is the large number of potential users and services that demands a scalability and agility impossibly provided by conventional models (like point-to-point connection of clients and servers). Finally, a strong requirement in educational scenarios is the acceptance and effective learning outcome by the users, not only those with a less technical background, which requires an intuitive and satisfying client-side interface. Here, the bow is drawn back to the desired tangibility and somatic, sensual perception of rich interaction spheres, which we tried to transfer from physical to virtual environments.

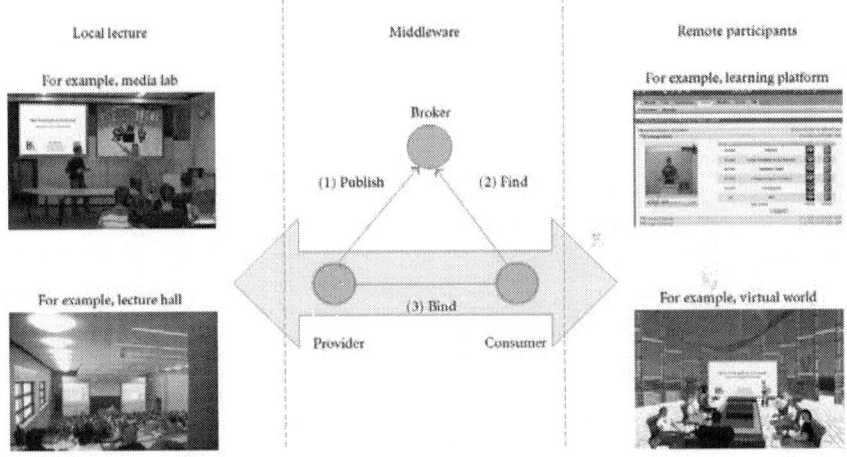

Figure 1: A Service-Oriented Architecture transparently interconnects different service providers and consumers (here: for telelecturing) without requiring the clients to have specific knowledge on given infrastructure and protocols.

Beyond simple brokerage, our service-based middleware offers some unique features which we would like to point out here.(i)Different network technologies are bridged transparently [27] (like Ethernet, WiFi, Bluetooth, and ZigBee), since there are many specific devices and networks in an educational context that have to be involved.(ii)Moreover, interoperability on service level is realized by an abstract layer for unifying service publication, brokerage, and use [27]. This was necessary because of the increasing number of service technologies.(iii)Finally, context awareness both on network and on service layer is integrated [13] in order to adapt the system behaviour to the current situation of the users, their environment, and the system itself.

That is why we called this central instance not only a broker, but a University Service Bus—indicating that there are complex functions performed by the middleware on behalf of the other system components (and finally of the users themselves).

3.2. Interactions in Virtual Worlds

From the users' point of view, there are three levels to deal with a digital system [28], which can be also applied to a virtual 3D world.(i)They may stay passive without influencing their environment. This includes to move around (e.g., walk, drive, or fly), to examine persons or objects (like exhibits in a museum), or to consume presented content (e.g., textual or audiovisual elements).(ii)They may actively shape their environment according to their own visions. Typical possibilities are presentation of any content (like presenting a poster or giving a talk) and creation or modification of certain elements (e.g., 3D models or simulations).(iii)They may interactively communicate with other users (e.g., by text or voice chat) and objects (not necessarily just in the virtual world itself but maybe also in external environments).

In general, the degree of intensity is rising from level to level. As a specialty of virtual 3D worlds, navigation, and interaction are similar to our actions in reality and thus are perceived to be more simple, natural, and intensive [5] than traditional computer-based interaction patterns. As a consequence, a particular suitability of virtual 3D worlds for teaching and learning can be seen. Typical application scenarios from an educational point of view are the following:

- distribution of previous lectures (in terms of a slide show or video) for passive, asynchronous reception, similar to a PodCast [29],

- synchronous transmission of video data and moves of an avatar, similar to a virtual video conference [30],

- additional discussions and/or reflections among students or with the tutor [27].

As far as we know, existing scenarios are restricted to a single virtual world up to now. In principle, a connection to other environments (virtual or physical) is possible, too. We built an infrastructure that fulfills all of the above-mentioned tasks. Primarily, ongoing face-to-face lectures are provided as a service and can thus be invoked by any platform in real-time. For Second Life, this is realized by presenting the slides and annotations of the lecturer on the virtual canvas, and by mapping the lecturer's voice to the avatar in the virtual world. Also, additional sounds can be mapped to the virtual lecture hall. Secondarily, all lectures are recorded and stored in an archive. In case that there is no ongoing live event, these recordings can be accessed from Second Life, too. For the users, there is no difference to be seen between this asynchronous playback and a live transmission except missing features for interaction. These interactions are the third field of our developments. We support a transparent, personal communication between users in the virtual world and onsite. Again, all these features are not limited to a special lab on the campus or to Second Life as virtual counterpart. The broker architecture allows for a highly flexible and dynamic deployment/invocation of services no matter from their origin and the targeted platform. The only prerequisition is a registration of the event at the broker, carried out by the lecturer prior to the presentation.

The added value a virtual world provides in contrast to conventional face-to-face teaching is not only to copy a classroom or lab setting and to broadcast a lecture in the Web. Of course, this scenario is important especially for inexperienced users—teachers as well as students—in order to orientate themselves. The advantage is the almost unlimited changeability of the environment. From an educational point of view, this prevents the teacher from circumstantially explaining unknown situations (e.g., for exploring foreign cultures) or tedious theory (e.g., traffic rules for a driving license)— an appropriate scenario can be created for the students. From a psychological point of view, this requires the users to deal with a potentially instable environment. Moreover, the efforts for the teacher shall not be underestimated. Providing well-known, unchangeable conditions and locations (like a virtual lecture hall with a connection to real-life settings) as a framework to embed and to experience alternating scenarios may help them to cope with this demand.

Another important point is communication. The platform offers possibilities to get in contact with a huge number of individuals in an informal way. In contrast to traditional learning platforms (with email, forums, chats, whiteboards, and so on), the 3D modelling of objects, persons, and their behaviour creates a kind of social presence. The apparent visibility and tangibility of objects and persons makes it easier for users to (inter)act in an unconstrained and natural way. Nevertheless, interaction always requires a counterpart, which in turn requires a given organizational structure (like lecture or consultation hours) to tackle the problem of lonely avatars in an almost unlimited space. We focus the combination of presentation and communication processes known from a classroom or lab with those in virtual worlds (here, Second Life). This implies a close correlation between appearance and behaviour of persons as well as objects in both environments. Tutor and students can individually choose to join the combined scenario either from the face-to-face or the virtual setting. Moreover, mutual control on the ongoing lecture and the media equipment, respectively, is possible from both environments.

3.3. System Interfaces

We had to consider the interfaces of existing systems in order to develop our framework, namely, the media equipment and control centre in our lab and the virtual world. The challenge was to interconnect these worlds using the SOA without touching regular operation of these systems.

The media control centre onsite acts as a provider of the media services. It contains an audio and video crossbar switch, which assigns all available inputs/outputs. This is controlled using a wireless panel via the internet control system protocol (ICSP). Instead of this panel, our service wrapper sends such ICSP commands to the switch, and thus is able to access and control the media streams even from remote. Other services have been developed for audio/video streaming [13]. This includes live streams of ongoing lectures as well as ondemand streams of archived e-lectures from a central media server. Though the internal realization is completely different, the use is fully transparent for the consumer. The user may only recognize additional interaction possibilities for live streams. We use a Darwin streaming server to broadcast streams, since it offers various data formats and simultaneous client connections, in contrast to other solutions.

Second Life is a complex virtual world model that is completely hosted, simulated, and rendered on servers of the vendor Linden Lab. Though there are some possibilities for own extensions in the dedicated Second Life client software, this does not apply to server-side communication mechanisms that have to be facilitated for realization of an SOA. That is why we introduced a so-called surrogate as a transmitter between available services and Web Service protocols on the one hand and SL's native communication mechanisms on the other hand. With the help of the Linden scripting language (LSL) 3D objects can been extended by interactive functions [31] for HTTP connection to the surrogate and thus to the SOA. Events can be triggered by mouse clicks, timers, or messages on the SL-internal communication channels.

Moreover, we intended to integrate mobile devices of onsite students (cell phones) in order to enable personal communication with virtual participants. Here, we had to bridge heterogeneity on network and service level (Bluetooth and its services on the phones, and IP and Web Services in the Internet). We developed a so called general purpose access Point (GPAP) to tackle this problem; this core element of our infrastructure lies outside the focus of this paper.

3.4. Realization of the 3D User Interface

The design of a Second Life interface to an SOA covers three major aspects. First of all, the overall appearance must be appealing and—for better acceptance—without an explicit reference to E-learning. Associations with an object or region from real-life create curiosity and may lead to a better identification. They are feasible to express local affiliations in virtual worlds that are assembled from a diversity of different countries and regions. For this reason, we decided to model two famous landmarks at the beach of Rostock Warnemünde: the historic lighthouse and the so called "Teepott" with its remarkable roof (Figure 2).

(a)

(b)

Figure 2: The lighthouse and "Teepott" have been transferred from real-life (as a nautical and touristical landmark) to our virtual site in Second Life (as a learning and communication space).

The second aspect concerns the interior of the buildings. It should reflect the purpose of buildings and elements and should encourage an (inter)active participation. We modelled a virtual media lab with table, several chairs and a canvas to build an open learning and communication environment (Figure 3).

Additionally, all internal functions must be distinguishable and intuitionally controllable. The service-based communication with other environments should occur seamlessly. We use an access model that is based on the Second Life group model: Registered members of our group are considered as trustable avatars and are allowed to control the virtual and real equipment. They can also authorize other avatars for specific events or lectures. Guest avatars can only consume content and communicate with other students or the lecturer by text and voice chat.

(a)   (b)

Figure 3: The main equipment of the media lab as well as additional elements for room decoration and interaction is modelled in Second Life, after all providing a much more attractive atmosphere than the original lab.

3.5. Exemplary Use Cases

From the many functions the virtual environment can fulfill we would like to explain four in more detail:

- controlling media equipment in the onsite lab from the virtual world,

- playing archived lectures in the virtual world,

- participating in ongoing lectures from the virtual world,

- communicating with other students across both worlds.

The media equipment in the onsite lab is controlled by a so called head-up display (HUD) which appears when an avatar touches a dedicated 3D object. We chose a keyboard as 3D object to combine familiar functionality with the aesthetics and beauty of personal media [32]. It acts as an interface to the media control service provided by the onsite lab. The HUD dialog shows the different sources and drains for multimedia signals (audio and video), and the user can assign signal routings. This is depicted in Figure 4. Technically, this is realized by invoking the media control service which transmits the commands to the lab.

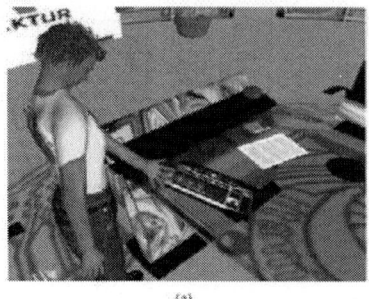 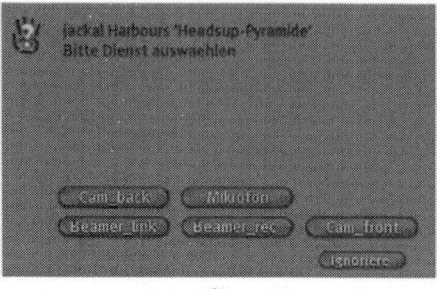

(a)   (b)

Figure 4: As an additional 3D element that attracts the user's attention a keyboard encapsulates interaction with invisible objects and non-3D, dialogue-based interaction (e.g., for controlling the technical equipment of the real-life media-lab).

The main use case the system was designed for is streaming of lecture recordings, which can take place live and ondemand. Lecture streaming from archive is a very simple scenario regarding both implementation of the system and interaction in the virtual world. An intuitive touch on the canvas of the virtual world [33] lets an avatar open a dedicated HUD for directly selecting from available video sources to be displayed on the canvas. The broker provides a list of currently available streams in the system, and the user selects from this list. Afterwards, the URL of this streaming service is mapped to the 3D model of the canvas, and playback begins. Depending on the media configuration, this contains only the slides (including digital annotations) or a combination with the lecturer's video. Simultaneously, the voice of the lecturer is mapped to the virtual speakers. Here, the user stays passive throughout the lecture and is just consuming the multimedia content. However, the learning effect is the same as with traditional lecture recordings—but playful students may find it more attractive to follow them in a lab-like setting than in a classical web browser. However, the immanent benefits of the virtual world do not have an effect here.

An enhanced use case is live streaming of an ongoing lecture. Typically, this is associated with the lecturer being present both in the classroom and in the virtual world (by means of an avatar). He starts recording and live streaming onsite, which is recognized by the broker. Thus, the stream is automatically available for playback on the virtual canvas. We experimented with different onsite configurations and considered a scenario with two screens (one for the slides of the lecture and one for depicting the virtual world—please refer to the photograph in Figure 3) to be the best solution. Now, all onsite and virtual participants can directly follow the lecture as if they were really together, which offers an added value compared to the more isolated ondemand settings. Social presence and intuitive interaction with the system create a comfortable atmosphere, which motivates the students to play a more active role and thus to intensify the process of knowledge acquisition.

There are also some possibilities for interactivity, to a limited extent. First, the Second Life (text or audio) chat will directly reach onsite participants if the virtual world is projected onsite. Audio messages can be handled in the same way as requests to speak by the audience onsite. But, we experienced that text messages can produce a high cognitive load for lecturers who have to be aware that there may appear comments or questions behind their back on the screen. Second, interactions with the students onsite need to find a way to the virtual world. We experimented with two mechanisms: The camera and microphone can be switched from the lecturer to the audience, which results in some delay of the lecture, and it is better handled by an additional technician than by the lecturer. Alternatively, the lecturer may simply repeat any question of the audience in order to transmit it to the virtual world. This does not require much effort. Nevertheless, we found these interaction possibilities insufficient.

That is why we additionally developed an innovative cross-technology communication concept [27]. We implemented a chat service running in the learning environment onsite. A messaging client on the users' mobile phones enables them to find other users by service discovery. Second Life also provides a chat interface to its users. We developed a virtual cell phone as messaging object which is able to find other messaging objects. If an avatar is "wearing" this object, it connects to the messaging service in the lab onsite and requests for the list of available messaging partners. Furthermore, it registers itself as messenger. Both in Second Life and on the Bluetooth phones the group of available chat partners is displayed as list, and the users (or their avatars, resp.) are able to send messages to all chat partners no matter where they are (Figure 5). This is the basis for collaborative settings beyond traditional instruction. In the general pedagogical model as well as in our specific scenario, the teacher steps back acting rather like a trainer or moderator. Such arrangements have proven to be suited for in-depth learning experiences resulting in highly transferable and applicable knowledge.

Figure 5: The cross-platform message exchange bridges between a mobile phone (Bluetooth) and the Second Life chat functions (Ethernet and Web Services). Students make use of their personal phone or corresponding virtual devices.

4. EVALUATION OF THE SYSTEM

The prototypical solution of the presented system was developed in a course on Web 2.0 and Second Life. For the students, this was associated with a wide range of actions that took place in the physical as well as the virtual environment: listening to introductory talks, preparing and giving their own talks on advanced topics, discussing these topics, extending the conception, implementing the virtual environment itself, testing and presenting the results of their work. We evaluated the infrastructure in three subsequent master courses with all together 25 students and different lecturers; two with traditional instruction and one with a game-based setting [34]. All talks have been recorded and deployed in Second Life as well as in our local learning platform, both live and ondemand. We are now going to describe our experiences from these courses with a focus on the virtual domain.

4.1. Technical Aspects

The frequency of using the service-based dissemination of lectures did continuously rise during the tests (compared to traditional linking of material in an E-learning platform) though students initially signalled no willingness to make use of these mechanisms. Comparing different transmission techniques, streaming was recognized as appropriate especially for participating and frequent reworking of a lecture, while download was considered as helpful mostly for targeted revision of lectures prior to exams.

As Figure 6 shows, the technical solution was rated mostly average to good. In general, the lecture integration into Second Life (right) polls worse than the compared version for the learning platform (left). As weak points, mainly the image and sound quality as well as the interaction facilities were mentioned. This helped us to fine-tune the resolution and sampling rate in the streaming server and to define some guidelines for a lecturer how to deal with annotation features of the recording software and with interactive elements in the pedagogical setting. Moreover, the students provided a number of suggestions for improvement especially of the Second Life solution and the overall organization of the lecture.

After all, we asked the students if they would make use of such offers during their studies, again. Regarding lecture streaming in the traditional learning platform 75% said yes and 25% perhaps. Regarding lecture streaming in Second Life, 50% said yes, 25% said perhaps, and 25% said no. Our conclusion is that using the virtual world rather makes sense in highly interactive settings like project-based or game-based learning.

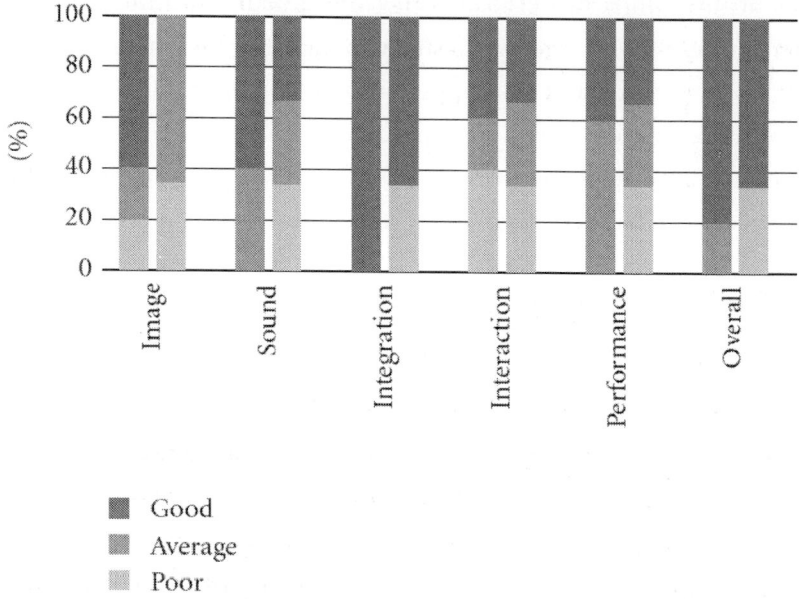

Figure 6: Student ratings for the quality of integrating onsite lectures into a learning platform (left) and a virtual world (right) showed relevant points for further improvement of our systems.

4.2. Social and Pedagogical Aspects

In an initial inquiry, we asked the students for their personal attitudes towards innovative E-learning materials and settings. The answers ranged from open minded (15%), to deliberate (35%), to sceptical (50%)—in contrast to conventional slides and scripts which were generally rated as important. The final attitude after finishing our tests changed, where lecture recordings gained the same relevance as slides and scripts (almost 100%), and also live transmissions and virtual lectures were rated as mostly important (75%). This demonstrates that the results of our evaluation were not biased by a general affinity of the students towards E-learning mechanisms or material.

Our subjective observations from a social point of view can be divided into three main phases of the course:

- In the first phase lasting a few weeks, the natural instinct of participants to play around in the virtual world dominated their behaviour. To a certain extent this also affected the instructors. For instance, many students experimented with the optical appearance of their avatar, widely exceeding limits given in real-life by biology, culture, or personal concerns. Some even acted in a more aggressive way, for example, tried to change, remove, or destroy virtual objects. This was encouraged by the prototypical nature of the technical realization as well as their feeling to be unobserved resulting from a lack of organizational structure in the new environment compared to traditional classroom or lab settings.

- In a second phase that lasted till the end of the course, we saw a significantly increased productivity of the students. They started to build up a team rather than a number of individuals, surely as a result of successful interaction and cooperation. Their identification with the project (visible, for instance, in dedicated logo shirts they designed and exchanged) seemed to be much higher in the virtual than in the physical environment, which we explain with the given potential to personally act out. Here, the innovative interaction sphere is a major benefit to exploit the students' individuality and creativity.

- Finally, even a third phase taking place after finishing the course was identified. Students continued to work and play with the system even beyond their official schedule. The team kept meeting physically as well as virtually, further refining the system, and there existed a strong interest to continue work in consecutive courses or projects which we are familiar only from other "tangible" projects, for example, mobile robots or field trips.

From our subjective perspective, the average learning outcome (in terms of given marks in the exams as well as in terms of obtained personal competencies of the students) was higher than in previous courses on similar topics. Unfortunately, because of the small number of participants in our master courses, we could not divide the students in test group and control group in order to quantify an increased learning outcome. However, some qualitative statements from students provided after the courses may validate our observations. In general, students reported that the interconnection of physical and virtual settings was "a valuable add-on to traditional lectures", and that "theory was more aligned to practice". One student involved in the game-based project confirmed: "Virtual projects in Second Life are a good idea and very exciting. It's annoying to spend more time with peculiarities of the SL engine than with the course topic itself. However, this is exactly what we as computer scientists have to cope with, so it's interesting to try that." Another one commented: "It was amazing to actually experience the results of my programming in a virtual environment. This was definitely worth the efforts." These observations are consistent with other experiments and studies in this field [35].

Our conclusion is that tangibility—no matter if physical or virtual—helps to foster identification with the project and intrinsic motivation of the students. This results in a better performance in single tasks as well as in increased self-confidence. This is in turn beneficial for the learning outcome, which is also compliant with other studies [36, 37]. In addition, well-known spatial as well as temporal conditions facilitate orientation and satisfaction of users in the virtual world.

We feel that these findings from our experiments as well as from educational research in general confirm the efficacy of our system. However, the contribution of our work is rather technical, a systematical approach to interconnect not only different tools, but also different teaching/learning settings which were isolated before.

5. SUMMARY AND FURTHER WORK

The service-based linking of face-to-face learning scenarios and different virtual learning environments described in this paper goes far beyond previous approaches. Connections between Second Life and traditional E-learning platforms already exist, but they are restricted to cross references or a common database. There are no systematic approaches to combine synchronous and asynchronous learning processes of both paradigms.

The presented system achieves a flexible and systematic coupling of platforms and tools of computer-aided teaching and learning in classrooms and virtual worlds. It consequently makes use of a service-oriented architecture (SOA). An intermediate service layer between the different environments contains all services that are provided by the specific platforms. Each environment is furthermore able to consume services available in this layer. For the first time, learner and lecturer can shape the specific learning and teaching processes in an ad hoc manner beyond predefined phases (Blended Learning) or environments (decoupled face-to-face and virtual learning processes). The individual arrangement can be modified during the lecture. In addition, the emerged independency allows a unification of synchronous and asynchronous learning and teaching scenarios. This can be easily realized across different educational providers. An interference of administrative areas of responsibility is not longer required thanks to the transparent encapsulation in an SOA.

Besides several courses at the University of Rostock, we also used the developed system for events like virtual online conferences or for remote (i.e., distributed) defence of students' theses, to name just two examples.

Nevertheless, the prototype can be extended at several points. A service-based feedback channel from the virtual learning environment into the presence learning environment does not yet exist. Although currently not required (clients and browser satisfactorily perform this task) it is desirable with regards to higher flexibility. A direct integration of SOA mechanisms into virtual worlds like Second Life would also increase the flexibility of the developed system, just like an extension of virtual environments to further data formats (e.g., PDF or Flash).

Furthermore, the interaction between teachers and learners can be designed more intuitionally. This is feasible by a fusion of the presented approach with principles and technologies of self organization and pervasive computing [38].

Finally, an advanced state of the system will allow us to transfer the scenario from Master to Bachelor courses (in order to attract more students and thus to gain a broader statistical base). We are confident that the service-based interconnection of virtual 3D worlds and real-life locations is an excellent interaction concept for different E-learning communities.

After all, the most obvious benefit of virtual worlds (to freely create and modify a 3D environment, for example, for simulations and role playing games [34]) is yet rarely utilized for learning activities. Though technical preconditions are given, this will require additional efforts from social and educational sciences as well as from media theory and design.

ACKNOWLEDGMENT

The authors are thankful to the German Research Foundation (DFG) for partially supporting this work within the Research Training Group "Multimodal Smart Appliance Ensembles for Mobile Applications".

REFERENCES

1. W. Coy, "Kulturtechnik Informatik," Informatik Spektrum, vol. 31, no. 1, pp. 30–34, 2008.

2. D. Tapscott, Growing Up Digital: The Rise of the Net Generation, McGraw-Hill, New York, NY, USA, 1998.

3. M. E. Gordon, N. Palakshappa, and D. J. Rowe, Real Life Brands in Second Life, Market Truths Limited, 2007.

4. K. Merrick, "Modeling motivation for adaptive nonplayer characters in dynamic computer game worlds," Computers in Entertainment, vol. 5, no. 4, article 5, 2008.

5. J. Cross, T. O'Driscoll, and E. Trondsen, "Another life: virtual worlds as tools for learning," eLearn Magazine, vol. 2007, no. 3, 2007, http://www.elearnmag.org/index.cfm#x26;article=44-1.

6. S. Turkle, "Always-on/always-on-you: the tethered self," in Mainstreaming Mobiles: Mobile Communication and Social Change, J. Katz, Ed., MIT Press, Cambridge, UK, 2008.

7. D. De Kerckhove, The Skin of Culture: Investigating the New Electronic Reality, Sommerville House, Toronto, Canada, 1995.

8. J. Truckenbrod, "Touchware," in Proceedings of the International Conference on Computer Graphics and Interactive Techniques (SIGGRAPH '98), p. 2, ACM, 1998.

9. G. Lakoff, Philosophy in the Flesh. The Embodied Mind and Its Challenge to Western Thought, HarperCollins, New York, NY, USA, 1999.

10. H. Ishii and B. Ullmer, "Tangible bits: Towards seamless interfaces between people, bits and atoms," in Proceedings of the 1997 Conference on Human Factors in Computing Systems, CHI, pp. 234–241, March 1997.

11. P. Lehsten, R. Zender, U. Lucke, and D. Tavangarian, "A service-oriented approach towards context-aware mobile learning management systems," in Proceedings of the 8th IEEE International Conference on Pervasive Computing and Communications Workshops (PERCOM '10), pp. 268–273, IEEE, April 2010.

12. R. Zender and U. Lucke, "Integrating context awareness with SOA: case studies for service-oriented and context-based architectures," in Proceedings of the 23rd International Conference on Architecture of Computing Systems (ARCS '10), VDE Verlag, Berlin, Germany, 2010.

13. U. Lucke, "Context-aware Multimedia Services for Mobile Lecture Streaming," in Multimedia Services and Streaming for Mobile Devices. Challenges and Innovations, IGI-Global, Hershey, Pa, USA, 2011.

14. C. J. Bonk and C. R. Graham, Handbook of Blended Learning: Global Perspectives, Local Designs, Pfeiffer, San Francisco, Calif, USA, 2005.

15. R. Mertens, N. Birnbaum, M. Ketterl, and R. Rolf, "Integrating lecture recording with an LMS: an implementation report," in Proceedings of World Conference on E-Learning in Corporate, Government, Healthcare, and Higher Education (E-Learn '08), AACE, Charlottesville, Va, USA, 2008.

16. C. Zhang, Y. Rui, J. Crawford, and L. W. He, "An automated end-to-end lecture capture and broadcasting system," ACM Transactions on Multimedia Computing, Communications and Applications, vol. 4, no. 1, article 6, 2008.

17. D. Burden, Datascape, Daden Ltd, Birmingham, UK, 2010, http://www.daden.co.uk/pages/datascape.html.

18. J. W. Kemp, D. Livingstone, and P. R. Bloomfield, "SLOODLE: Connecting VLE tools with emergent teaching practice in Second Life: Colloquium," British Journal of Educational Technology, vol. 40, no. 3, pp. 551–555, 2009.

19. G. Turban and M. Mühlhäuser, "A framework for the development of educational presentation systems and its application," in Proceedings of International Workshop on Educational Multimedia and Multimedia Education (EMME '07), pp. 115–118, September 2007.

20. M. Schmid, J. Stynes, and R. Kröger, "A distributed architecture for learning management systems supporting institutional collaboration," in Proceedings of World Conference on Educational Multimedia, Hypermedia and Telecommunications (Ed-Media '04), AACE, Charlottesville, Va, USA, 2004.

21. F. Trinta, C. Ferraz, and G. Ramalho, "Middleware services for pervasive multiplatform networked games," in Proceedings of the 5th ACM SIGCOMM Workshop on Network and System Support for Games (NetGames '06), ACM, October 2006.

22. S. Bergsträsser, T. Hildebrandt, L. Lehmann, C. Rensing, and R. Steinmetz, "Virtual context based services for support of interaction in virtual worlds," in Proceedings of the 6th ACM SIGCOMM Workshop on Network and System Support for Games (NetGames '07), pp. 111–116, ACM, New York, NY, USA, September 2007.

23. A. S. Tanenbaum and M. van Steen, Distributed Systems. Principles and Paradigms, Prentice Hall, Upper Saddle River, NJ, USA, 2nd edition, 2006.

24. C. Larman, Applying UML and Patterns: An Introduction to Object-Oriented Analysis and Design and Iterative Development, Prentice Hall, Upper Saddle River, NJ, USA, 3rd edition, 2004.

25. U. Lucke, "An algebra for multidimensional documents as abstraction mechanism for cross media publishing," in Proceedings of the 2nd International Conference on Automated Production of Cross Media Content for Multi-Channel Distribution (AXMEDIS '06), pp. 165–172, IEEE, December 2006.

26. T. Erl, SOA Principles of Service Design, Prentice Hall, Upper Saddle River, NJ, USA, 2007.

27. R. Zender, E. Dressier, U. Lucke, and D. Tavangarian, "Pervasive media and messaging services for immersive learning experiences," in Proceedings of the 7th Annual IEEE International Conference on Pervasive Computing and Communications (PerCom '09), IEEE CS Press, 2009.

28. D. A. Wiley, "Connecting learning objects to instructional design theory. A definition, a metaphor and a taxonomy," in The Instructional Use of Learning Objects, pp. 3–29, AIT/AECT, Bloomington, Ind, USA, 2001.

29. S. Cass, "Podcast Picks," IEEE Spectrum, vol. 44, no. 4, p. 53, 2007.

30. R. Nesson, Virtual Worlds, Harvard University, 2008, http://www.eecs.harvard.edu/~nesson/e4/.

31. P. Hiller, Design und Programmierung in Second Life, Franzis, Poing, Germany, 2007.

32. F. Heinrich, "The aesthetics of interactive artifacts—thoughts on performative beauty," in Proceedings of the 2nd International Conference on Digital Interactive Media in Entertainment and Arts (DIMEA '07), vol. 274, pp. 58–64, ACM, Perth, Wash, USA, 2007.

33. K. Mullet and D. Sano, Designing Visual Interfaces: Communication Oriented Techniques, Prentice Hall, Upper Saddle River, NJ, USA, 1994.

34. R. Zender, U. Lucke, D. Maciuszek, and A. Martens, "Interconnection of game worlds and physical environments in educational settings," in Proceedings of the 8th Annual Workshop on Network and Systems Support for Games (NetGames '09), Paris, France, November 2009.

35. D. H. Lim and M. L. Morris, "Learner and instructional factors influencing learning outcomes within a blended learning environment," Educational Technology and Society, vol. 12, no. 4, pp. 282–293, 2009.

36. P. Brauner, T. Leonhardt, M. Ziefle, and U. Schroeder, "The effect of tangible artifacts, gender and subjective technical competence on teaching programming to seventh graders," in Proceedings of the 4th International Conference on Informatics in Secondary Schools—Evolution and Perspectives (ISSEP '10), pp. 61–71, Springer, Berlin, Germany, 2010.

37. S. Do-Lenh, P. Jermann, S. Cuendet, G. Zufferey, and P. Dillenbourg, "Task performance vs. learning outcomes: a study of a tangible user interface in the classroom," in Proceedings of the 5th European Conference on Technology Enhanced Learning (EC-TEL '10), Lecture Notes in Computer Science 6383, pp. 78–92, September 2010.

38. D. Tavangarian and U. Lucke, "Pervasive University—a technical perspective," in it—Information Technology, vol. 51, pp. 6–13, Oldenbourg, München, Germany, 2009.

CHAPTER 9

Study of the Assessment Criteria on e-Learning Websites

Kuei-Chih K.C. Chuang[1] and Mei Chuan Tsai[1]

[1] *Graduate Institute of Technological and Vocational Education, National Yunlin University of Science and Technology, , Sec.3, University Rd., Douliu, Yunlin, Taiwan, R.O.C*

Abstract

This study aimed at exploring and discussing cognizance of teachers and students toward construction of e-learning websites. It was evaluated and it developed five assessment indexes required as "Assessment Guidelines on e-Learning Websites." These five assessment indexes were "teaching material and the structure," "layout design," "interface design," "interaction design," and "establishment of system configuration." The development of survey questionnaires was based on the above five assessment indexes as well. In this study, several goals are achieved, for example, teaching resources could be augmented, quality of Web-based instruction could be improved, learners' time and efforts in Web-based learning could be saved, and effects of Web-based teaching and learning could be highlighted. In the end, a digital Taiwan can become possible when the Web-based instructions follow the assessment guidelines and prevail over the aggressive competition.

Keywords

Assessment guidelines of an instructional website, Internet, Instructional websites, Internet server end, Assessment guidelines

1. INTRODUCTION

The twenty-first century has been now called the Internet Age which is greatly affected by the various computerized multimedia technologies, such as the Internet. With the wide range of interactive computer and Web technologies, it tends to produce more effective methods of transferring skills and knowledge than the traditional lecture-style approaches. For example, the professors can integrate notes, graphics, diagrams, full-motion video segments, audio segments, and hyper link texts and materials into a comprehensive website as a cognitive and motivational tool to facilitate teaching and learning. In addition, the students can acquire and construct knowledge and skills by accessing their learning performance and take part in discussing with instant feedbacks [1].

In the way of online learning theory and application, teachers and students can get advantages from flexible teaching and learning, and make much progress step by step. Teachers easily get the feedback by students' on-line questions and reply the answers to students as soon as possible. Besides, setting up the instructional websites from teachers by professional knowledge and skills makes effective responses for the demands by students. Eventually, teachers and students get the win-win advantages for each other from such a valid and cognitive system. However, educational curriculum designers and faculty face great challenges and potentials by new teaching experiences. More and more researches have focused on teaching and learning characteristics, learning styles, and interactions between teachers and students about Web-based instructional setting. Nowadays, it has become the greatest concern for all the teachers on what the criteria should be and how to design and evaluate an effective and interactive Web-based instructional setting. We have seen in the cases presented that clearly articulated assessment strategies are vital to the effective design of online courses and programs. The peer assessment case demonstrates a program-level solution to the need to provide a tool that assesses professional skills in group-level and individual-level performance within an online context. McCracken, Cho, Sharif, Wilson, and Miller [2] made a conclusion that "it presented that clearly articulated assessment strategies are vital to the effective design of online courses and programs. The peer assessment case demonstrates a program-level solution to the need to provide a tool that assesses professional skills in group-level and individual-level performance within an online context." Therefore, the more requirement for e-learning environments grows in higher education, the more needs for website application of learning and educational assessment strategy theory to design, develop, and deliver in e-learning environments.

Graff [3] pointed out that "evaluation feedback from participants indicated that each online task was rated positively." And Smith [4] made a comment that "it is suggested that as instructors make the transition from traditional to blended/online instruction, they consider jettisoning the traditional essay requirement and replace it with some form of 'assignment essay/peer review' system such as the one described. Contemporary Learning Management Systems facilitate peer review and peer assessment approaches in ways that were not available in traditional offline education." Somehow, doing the assessment study should consider that there are numerous limitations on how to learn performance that is evaluated by assessment theories and skills. The patterns as online testing show are usually in the form of multiple choice questions without any essay type of learning assessment. Most of the reasons for offering multiple choice tasks in e-learning are for the sake of ease of implementation and ease of managing learners' replies.

The study purposes were to establish a set of criteria for assessing instructional websites by using instructors' and students' perception of current experiences in instructional websites specifically for undergraduate- and graduate- level courses in order to meet the following objectives: collecting and understanding the settings and operating rules of the present instructional websites, analyzing and developing the updated policies of the instructional websites, exploring the users' and designers' perceptions, and providing tips for building an effective instructional website. Web evaluation criteria include ideas for incorporating Web evaluation into the curricula that promote information literacy. In particular, we intended to explore five-section criteria related to the website evaluation. The five criteria were

(1) website material development,

(2) website graphic design,

(3) website interface,

(4) website communication interaction, and

(5) website system.

The study measures included collecting and reviewing reference papers related to instructional websites, analyzing and inducting the websites' advantages and disadvantages, interviewing experts, and analyzing evaluation of e-learning website. During the study period, the research targets covered 33 instructors who established their own instructional websites at the first semester of the academic year 2013 at National Yunlin University of Science and Technology, 35 teachers who participated in the teacher workshop of research methods for teachers of science and technology in south Taiwan, and 240 college students of the National Yunlin University of Science and Technology. From the participants' responses of the survey questionnaires, the most essential features and characteristics required for a quality instructional website were concluded as the assessment guidelines of an instructional website.

The expert group had revised the assessment guidelines of instructional websites twice, and the effectiveness of the guidelines was high. From the analysis of data collected, it was found that the items of indexes for the assessment guidelines of instructional websites were recognized by the research targets including navigation group, teacher group, and student group. Furthermore, these items were consistent with the hypothesis of the study and results of the analysis of pilot test data conducted by the navigation group. As a result, the guidelines which resulted from the five assessment indexes displayed much high value. The research results were also regarded as database where teachers of all disciplines could refer to whenever they would develop their own instructional websites.

2. METHODS

2.1. Participants

The samples of this study consisted of three groups. In the first group called control group for pilot test as well, there were 20 pre-teaching teachers who still were students enrolled in the Teacher Education Program at the National Yunlin University of Science and Technology (NYUST). In the second group, there were 33 professors who had built instructional websites for their courses at NYUST since academic year 2012, and for the other group, there were 45 teachers who attended "The Research Methods Conference of Southern Institute and College" held by NYUST in 2012. In the third group, there were 240 current NYUST students enrolled in the College of Engineering, College of Management, College of Design, and College of Humanities and Science.

2.2. Instrument And Return Rate

A descriptive survey method was used to collect data. After a close review of studies [5, 6, 7, 8] and particular website guidelines [9, 10, 11], a closed-ended questionnaire, called Criteria of Instructional Website, was designed to gather data related to participants' perceptions of experiences on instructional websites. The questions related to each of the five criteria are listed in Table 1 including six variables, gender, professional field, years of hands-on experience of Internet surfing, educational background, years of teaching (for faculty only), and purposes of using the Internet, also set up for analyzing vocational education faculty's and students' perceptions of instructional websites. The overall response rate for this study was 77.6 %.

3. RESULTS

3.1. Control Group (N = 20)

The number of pre-teaching teachers of the control group was 20 (n = 20). Among the participants, 75 % were females, 55 % with master degree, 30 % were studying in College of Management, and 70 % with experiences of

Internet surfing for less than four years. And they were all sharing similar purposes of Internet surfing, such as e-mails, chatting on-line, gathering information, on-line games, and on-line learning. In addition, the results indicated that the participants rated the hypothesized five criteria as important elements with an average of more than 3.4 of the five-point Likert scale on the same variables mentioned above.

TABLE 1. SAMPLE QUESTIONNAIRE

Item of criteria	Sample question
1. Website material development	The material should be genuinely related to the Web name that tells
2. Website graphic design	The instructional website design should contain multimedia effects, such as sounds, 3-D pictures
3. Website interface	The web should offer an exact time for downloading each file
4. Website communication interaction (interactivity)	The instructional website should offer some kinds of sections for motivating students to participate in discussion
5. Website system	The instructional website should offer a diversity of learning assessment

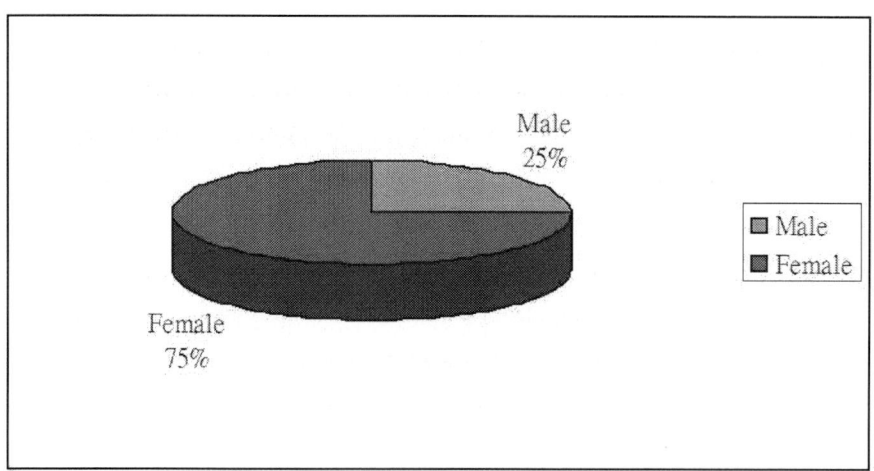

Figure 1. Distribution Of Gender Of Pre-Teaching Teachers

Among pre-teaching teachers who attended the seminar hosted by the National Yunlin University of Science and Technology Teachers' Center in

2012, the distribution percentages were as follows: male teachers with 25 % and female teachers with 75 %.

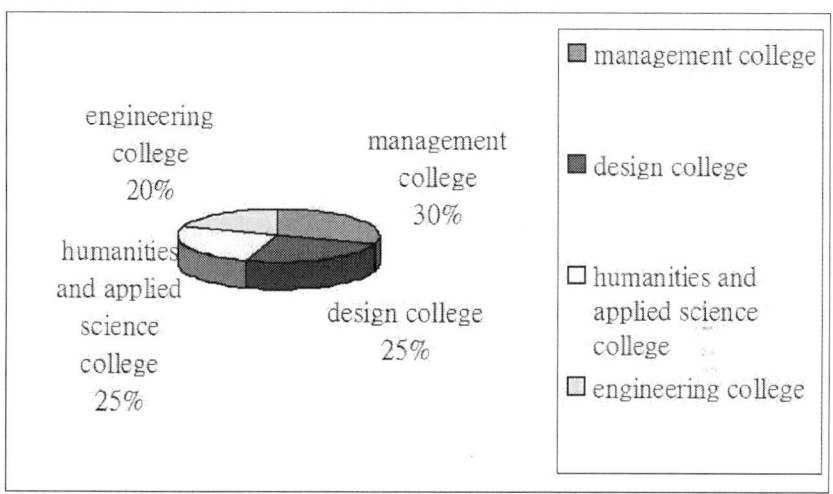

Figure 2. Distribution Of Pre-Teaching Teachers Who Studied In College

Among pre-teaching teachers who attended the seminar hosted by the National Yunlin University of Science and Technology Teachers' Center in 2012, the distribution percentages were as follows: College of Management with 30 %, College of Design with 25 %, College of Engineering with 20 %, and College of Humanities and Applied Sciences with 25 %.

Among pre-teaching teachers who attended the seminar hosted by the Center of Teacher Education in the National Yunlin University of Science and Technology Teachers' Center in 2012, the percentages were as follows: master degree with 55 %, four-year bachelor's degree with 20 %, two-year bachelor's degree with 20 %, and doctoral degree with 5 %.

Among pre-teaching teachers who attended the seminar hosted by the Center of Teacher Education in the National Yunlin University of Science and Technology Teachers' Center in 2012, the percentages were as follows: never used with 0 %, 1 year and less with 30 %, 2-4 years with 40 %, 5-7 years with 30 %, and more than 7 years with 0 %.

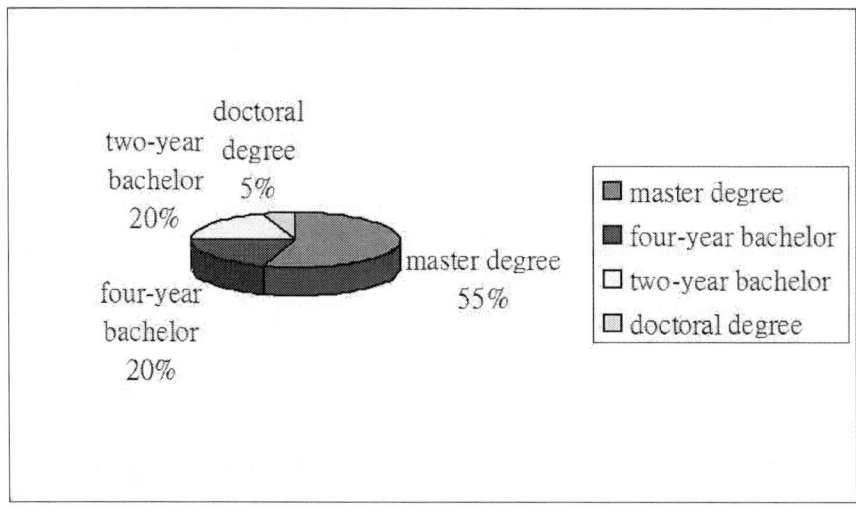

Figure 3. Distribution Of Pre-Teaching Teachers

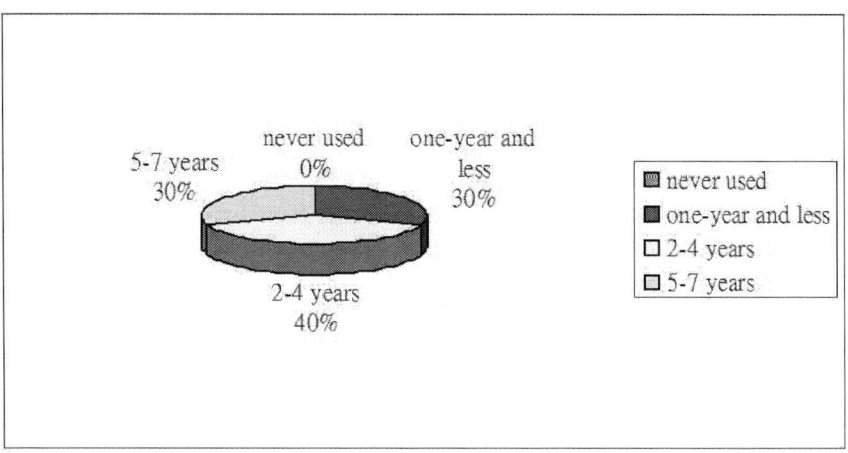

Figure 4. Distribution Of Time In Using Internet For Pre-Teaching Teachers

Among pre-teaching teachers who attended the seminar hosted by the Center of Teacher Education in the National Yunlin University of Science and Technology Teachers' Center in 2012, the percentages were as follows: receiving and sending e-mails with 25 %, BBS or online chatting with 20 %, browsing data with 25 %, playing online games with 25 %, and online learning with 5 %

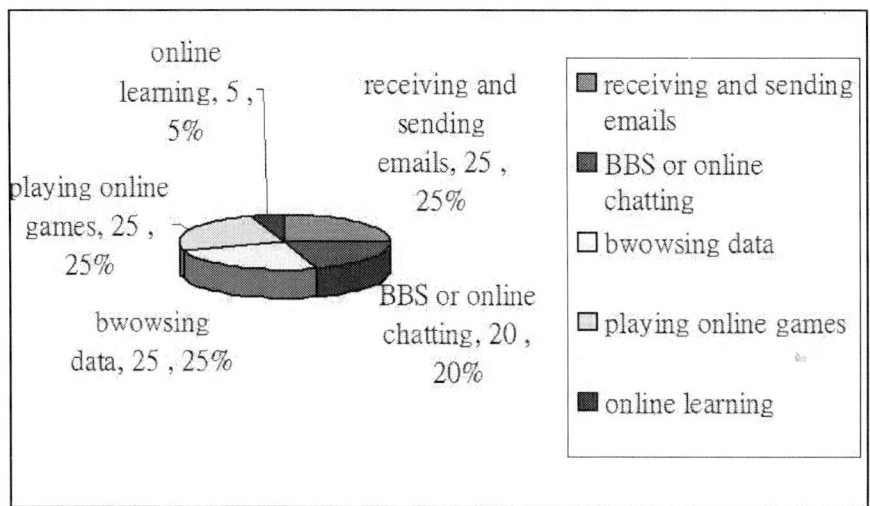

Figure 5. Distribution Of The Purposes Of Using Internet For Pre-Teaching Teachers

3.2. Teachers Group: (N = 78)

Two thirds of the participants were males teaching in the College of Management with 20.8 %, College of Art and Design with 20.8 %, College of Humanities with 20.8 %, and College of Engineering with 37.5 %. One fourth of them had less than 15 years of teaching experiences. Interestingly, a teacher who had more than 10 years of teaching experiences indicated that the main purpose of Internet surfing was only for e-mail sending/receiving rather than instructional purposes. Compared with others, most of the teachers indicated that they used Internet for class activities with 83.3 %, and information collection with 77.5 %. More importantly, there was not significant difference between the control group and the teacher group for the perception of the instructional website based on the hypothesized five criteria at the probability of 0.05. Nevertheless, different college faculties shared slightly different opinions about the importance of item 1 of the criteria of website material development (e.g. the material should be compatible with the title which it belonged to) as shown in Table 2. Also, different college faculties shared slightly different perceptions on the items of the appropriate location of frame design in the criteria of "website interface" based on the Schaffer test.

Table 2. Website Material Development, Item #1 Schaffer Test

College N	N	Alpha =0.05	
		1	2
Management	29	3.52	
Engineering	3	4.00	4.00
Art and Design	21	4.24	4.24
Humanities	6		4.83
		0.465	0.335

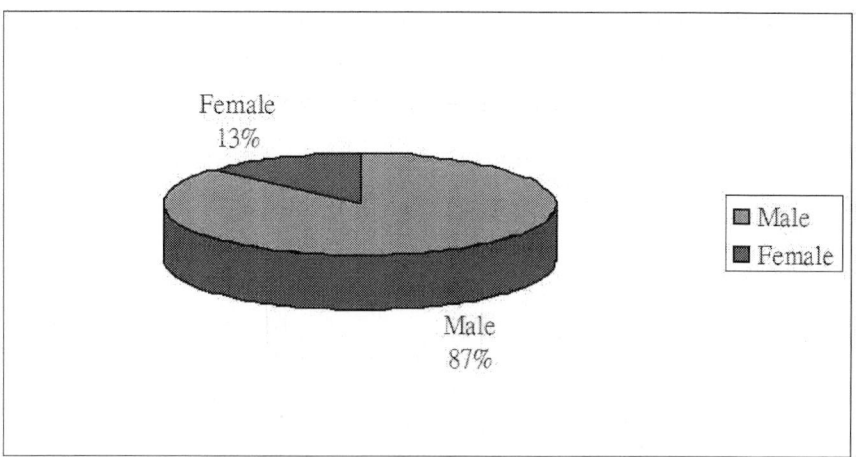

Figure 6. Distribution Of Gender Of Yunst Teachers Who Have Built Web Pages As Teaching Aids

Among YUNST teachers who have built Web pages as teaching aids, the percentages were as follows: male with 87 % and female with 13 %.

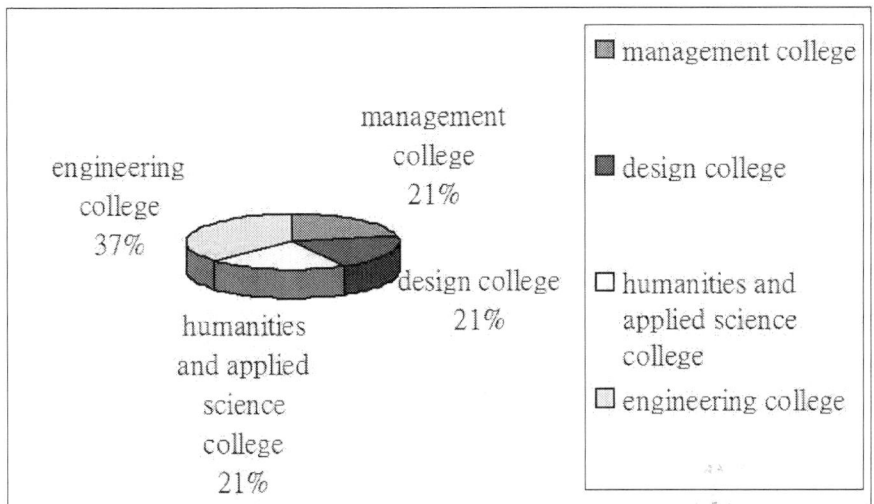

Figure 7. Distribution Of Yunst Teachers Who Have Built Web Pages As Teaching Aids

Among the college of YUNST teachers who had built Web pages as teaching aids, the percentages were as follows: College of Management with 21 %, College of Design with 21 %, College of Humanities and Applied Sciences 21 % and College of Engineering with 37 %.

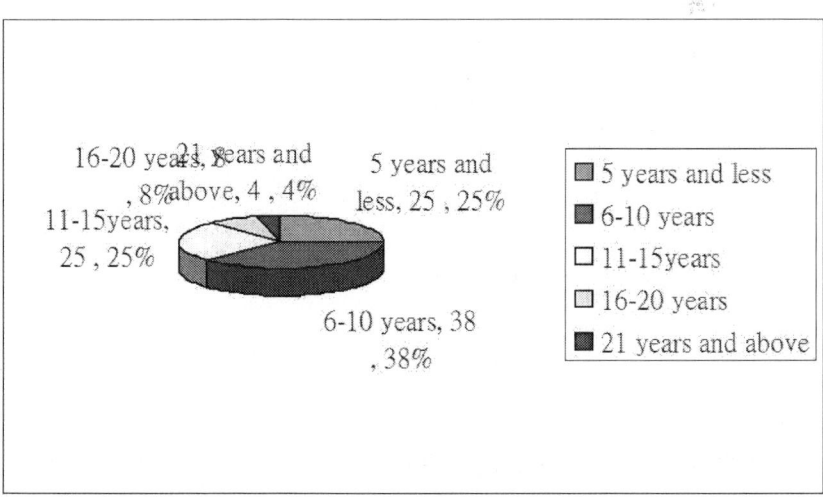

Figure 8. Distribution Of Seniority Of Yunst Teachers Who Have Built Web Pages As Teaching Aids

Among seniority of YUNST teachers who have built Web pages as teaching aids, the percentages were as follows: 5 years and less with 25 %, 6-10 years with 38 %, 11-15 years with 25 %, 16-20 years with 8 %, and 21 years and above with 4 %.

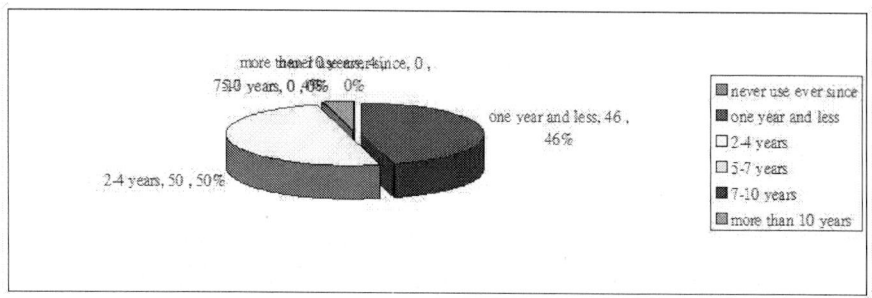

Figure 9. Distribution Diagram Of Time Yunst Teachers Who Have Built Web Pages As Teaching Aids Used The Internet

Among the time YUNST teachers who have built Web pages as teaching aids used the Internet, never used was 0 %, one year and less 46 %, 2-4 years 50 %, 5-7 years 0 %, 7-10 years 0 %, and more than 10 years 4 %.

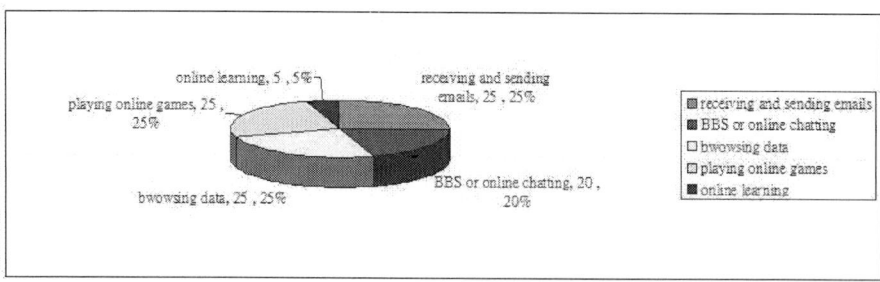

Figure 10. Distribution Program Of The Purpose Of Yunst Teachers Who Have Built Web Pages As Teaching Aids To Use The Internet

Among the purpose of YUNST teachers who have built Web pages as teaching aids to use the Internet, the percentages were as follows: receiving and sending e-mails with 25 %, browsing data with 25 %, BBS or online chatting with 20 %, playing online games with 25 %, and online learning with 5 %.

3.3. Nyust Students Group (N = 240)

About half of the respondents were male or female. Most of them studied in the College of Management (31.8 %) and College of Engineering (31.3 %) with a college degree (47.0 %), and two- to four-year Internet application experience (55.6 %). The results also revealed that the major purposes of surfing the Internet were for e-mail (69.7 %) and information collection (76.1 %). Based on the results of t-test analysis, there was no significant difference between the control group and the student group for the perceptions of the instructional website based on the hypothesized five criteria at the probability of 0.05. However, different college students have slightly different perceptions on the items of the appropriate updated relevant information given on the side of the site in the criteria of "website material development" based on the Schaffer test.

Third, NYUST students (n = 240), about half of the respondents were male/female. Most of them studied in the College of Management (31.8 %) and Engineering (31.3 %) with a college degree (47.0 %) and two- to four-year Internet application experience (55.6 %). The results also revealed that the major purposes of surfing the Internet were for email (69.7 %) and information collection (76.1 %). Based on the results of t-test analysis, there was no significant difference between the control group and the student group for the perceptions of the instructional website based on the hypothesized five criteria at the probability of 0.05. However, different college students have slightly different perceptions on the items of the appropriate updated relevant information given on the side of the site in the criteria of "website material development" based on the Schaffer test.

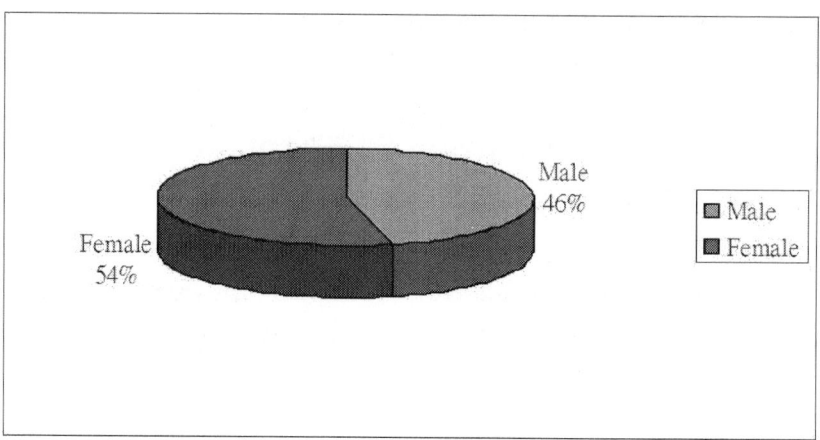

Figure 11. Distribution Of Gender Of Yunst Students

Among YUNST students, the percentages were as follows: male with 46 %, and female with 54 %.

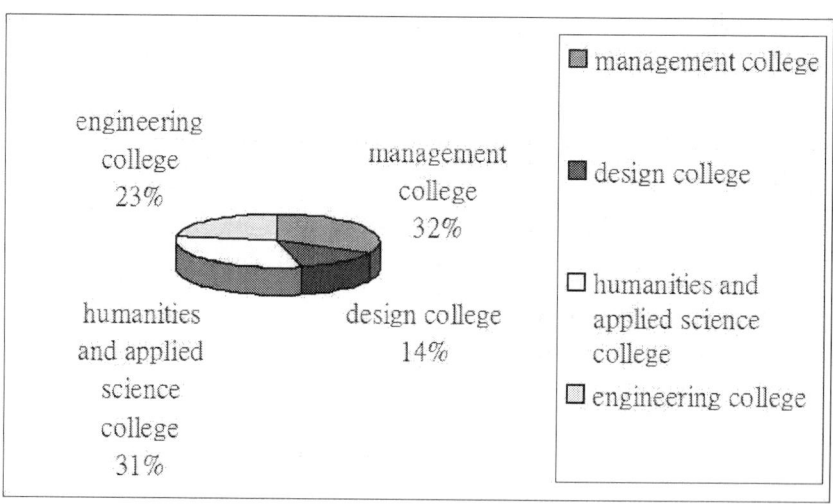

Figure 12. Distribution Of College Students At Yunst

Among students at YUNST, the percentages were as follows: College of Management with 32 %, College of Design with 14 %, College of Engineering with 23 %, and College of Humanities and Applied Sciences with 31 %.

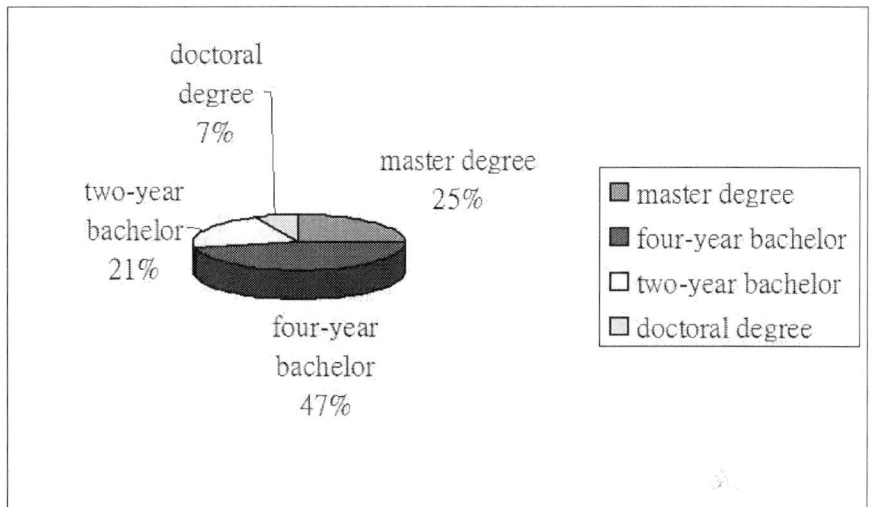

Figure 13. Distribution Of Students' Degree At Yunst

Among the degrees of YUNST students, the percentages were as follows: master's with 25 %, four-year bachelor's with 47 %, two-year bachelor's with 21 %, and doctoral with 7 %.

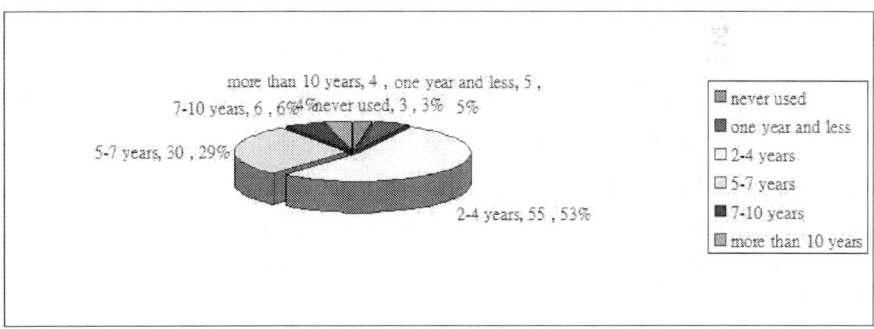

Figure 14. Distribution Of Years In Using The Internet Of Yunst Students

Among the years YUNST students have used the Internet, the percentages were as follows: never used with 3 %, less than 1 year with 5 %, 2-4 years with 53 %, 5-7 years with 30 %, and 7-10 years with 6 % and more than 10 year with 4 %.

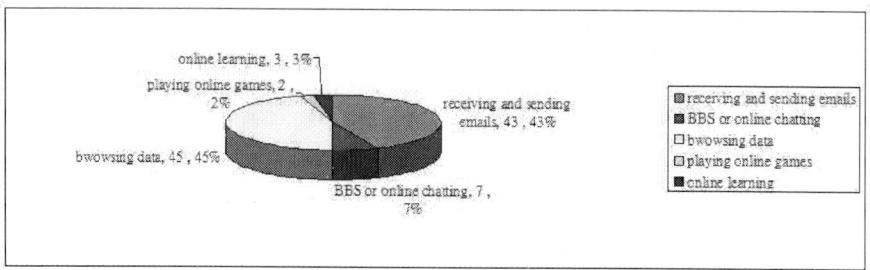

Figure 15. Distribution Of Purpose Of Yunst Students In Using The Internet

Among the purpose of YUNST students in using the Internet, the percentages were as follows: receiving and sending e-mails with 43 %, BBS or online chatting with 7 %, browsing data with 45 %, playing online games with 2 %, and online learning with 3 %.

4. CONCLUSION

The central concern of this study was to develop efficient ideas of instructional websites criteria for teachers and instructional Web designers to use and build excellent instructional Web environment. The results revealed that hypothesized criteria were perceived as a highly significant contribution. Indeed, the results could help Web designers construct highly effective and interactive instructional Web environment and bring up education in Taiwan into a higher-level digital learning situation. Based on the results, the study provided some recommendations for further instructional Web establishment and further studies as follows:

1. Instructors should be flexible in constructing their instructional websites specifically to the target students' educational backgrounds.

2. A comparative study for perception of long-term teaching experienced teacher versus new or short-term teaching experienced teacher on criteria of instructional website was suggested for further study.

3. Researches must go on to expand to other schools, faculty and students to deepen scopes and verify the findings.

5. ACKNOWLEDGEMENTS

The authors would like to thank the National Science Council and National Yunlin University of Science and Technology for the grant support.

The study has much appreciated the enthusiastic participants including three instructors who established their own instructional websites in the first semester of the academic year 2013 at the National Yunlin University of Science and Technology, 35 teachers who participated in the teacher workshop on research methods for teachers of science and technology in south Taiwan, and 240 college students of the National Yunlin University of Science and Technology.

REFERENCES

1. Kuo, C.H. & Chang, T.S. English On-line: Designing a Web-based EST Course. Proceedings of 2003 International Conference on English Teaching and Learning in the Republic of China. Providence University Taichung (2003).
2. McCracken, J., Cho, S., Sharif, A., Wilson, B., & Miller, J. "Principled Assessment Strategy Design for Online Courses and Programs." Electronic Journal of E-learning, 10.1, (2012):107-119.
3. Graff, Martin. "Cognitive style and attitudes towards using online learning and assessment methods." Electronic Journal of e-learning 1.1, (2003): 21-28.
4. Smith, Martin. "Can Online Peer Review Assignments Replace Essays in Third Year University Courses? And If So, What Are the Challenges?." Electronic Journal of e-Learning 10.1, (2012): 147-158.
5. Martinez, M. Successful Web Learning Environments: New Design Guidelines. ERIC Document ED 446745 (2000).
6. Ministry of Education. Education in the Republic of China. Bureau of Statistics, MOE: Taipei. (1996)
7. Ministry of Education. (2001). Education in the Republic of China. Taipei: MOE.
8. Pan, A. C. (1998). Optimize the Web for Better Instruction. ERIC Document 421098.

9. Technology for all Americans Projects. Technology for All Americans: A Rationale and Structure for the study of technology. Blacksburg: VA(1997)

10. Summary, R. & Summary, L. (1998). The Effectiveness of the World Wide Web as an Instructional Tool. ERIC Document 431393.

11. Wang, J. & Chuang, K.C. (2003). A Study of Instructional Website Guidelines. World Conference on E-Learn in Corp., Govt., Health, & Higher Ed., Vol. 2002, Issue 1, pp.999-1005.

CHAPTER 10

Considerations on Barriers to Effective E-learning toward Accessible Virtual Campuses

Salvador Otón[1], Héctor R. Amado-Salvatierra[2], José Ramón Hilera[1], Eva García[1] and Antonio García[1]

[1] *Universidad de Alcalá, Spain*
[2] *Universidad Galileo, Guatemala*

Abstract

Nowadays, the implementation of virtual campuses is a reality, both in academic settings and in the workplace. However, there are several challenges associated with the implementation of effective learning outcomes via e-learning. In this chapter in particular, the use of e-learning to reach students with disabilities and the barriers that they may have will be presented. In this sense, e-learning solutions adopted by several institutions are encouraged to validate and promote accessibility in a virtual campus. A large myriad of research related to accessibility in distance education systems is available in literature, and the most relevant studies and standards are presented in this chapter as a starting point for education institutions looking at improving the accessibility in their own virtual campuses. This work is intended to be relevant both to teachers and lecturers who use e-learning for their courses, and to those involved in the design, setup, and maintenance of e-learning systems, whether from a pedagogical or technical perspective to take into account the accessibility for students with disabilities. This work will explore on the accessibility of the basic stone of the e-learning process, the learning objects. An analysis of the IMS AfA v3.0 specification will be presented as a starting point to develop an accessible and adaptable online course, based on the student's preferences, within an accessible virtual campus.

Keywords

accessibility, learning objects, adaptability, disability, e-inclusion

1. INTRODUCTION

A virtual campus is an environment based on a web technology that provides facilities for the development, management, and publication of content that contributes to the process of teaching and learning. The process of teaching and learning enhanced by technology is commonly known as e-learning. The virtual campus is the fundamental element on which a virtual education project is based. If it is an accessible virtual campus, it must be ensured that all functionality can be used by any user, including users with disabilities.

There are several challenges associated with the implementation of effective learning outcomes via e-learning within a virtual campus. In this chapter in particular, the considerations on the use of e-learning to reach students with disabilities and the barriers [1–5] that they may have will be analyzed, providing the basic knowledge to prepare an accessible virtual campus.

This chapter is structured as follows: A state of the art on accessibility related to virtual campuses, highlighting studies related to the application of accessibility standards to improve the e-learning systems is presented in the first section. The first section explores on the main accessibility requirements for an e-learning campus. Then a review on the basic knowledge that the stakeholders involved in e-learning education should have in order to preserve and promote accessibility is presented. In particular, the authors propose an evaluation guideline on accessibility for virtual campus administrators. Finally, the considerations on the accessibility requirements of learning objects (LOs) are presented using the IMS Access for all v3.0 specification, the main objective of which is to simplify the definition of the accessibility metadata for learning objects and the preferences and needs of the users of these objects tracing them to students' related disabilities.

2. STATE OF THE ART ON ACCESSIBILITY RELATED TO VIRTUAL CAMPUSES

A virtual campus is an environment based on a web technology that provides facilities for the development, management, and publication of content that contribute to the process of teaching and learning. In this work, a virtual campus will be also referred as e-learning system and learning management system (LMS). In terms on legislation related to students with disabilities in e-learning, Edmonds [6] explored the different laws available and highlights the legal and technical concerns for education institutions. International legislation in terms of technological evolution related to e-learning is reflected on the Convention on the Rights of Persons with Disabilities (CRPD) in Article 9 (points 2.g an 2.h) [7]. The CRPD highlights the importance of promoting access to information and communications technology (ICT) for people with disabilities (PWD) and specially producing accessible content in early stages at minimum costs. Related to education, the (CRPD) in Article 24 recognizes the right to education. Countries that signed the CRPD must make sure that students with disabilities are able to get access not only to general education but also to tertiary education, vocational training, adult education, and lifelong learning without discrimination and on an equal basis with others.

In terms of accessibility, the International Organization for Standardization (ISO) defines accessibility as "the usability of a product, service, environment or facility by people with the widest range of capabilities" [8]. The World Wide Web Consortium (W3C), the organization in charge of developing web standards, created the Web Accessibility Initiative (WAI) with the aim of studying the problems of accessibility and propose solutions. One of its most known results is the Web Content Accessibility Guidelines (WCAG) 2.0 that establishes four principles that give the foundation of web accessibility: web content must be perceivable, understandable, operable, and robust [8].

In terms on learning objects accessibility, it is important to take into consideration the standard ISO/IEC 24751 [9–11] to describe the process of using an accessible online educational system, which takes into account the needs and preferences of the student and contains accessibility metadata of the learning objects. This chapter will explore also on the metadata for the learning objects using the IMS Access for All v3.0 specification [12], the main objective of which is to simplify the definition of the accessibility metadata for learning objects and the preferences and needs of the users of these objects.

2.1. General Requirements For Accessibility Of Learning Management Systems (Lms)

Learning management systems (LMS) are mainly based on web technologies through a client–server model, with an interface prepared to work base on HTML markup and presented in a web browser. For this type of systems, accessibility requirements should be followed, especially guidelines provided by the Web Accessibility Initiative (WAI) [13] part of the World Wide Web Consortium (W3C). These guidelines are summarized as follows:

- Authoring Tool Accessibility Guidelines (ATAG) [14]—guidelines intended to software used to create web sites and content

- User Agent Accessibility Guidelines (UAAG) [15]—addresses web browsers and media players, and especially related to assistive technologies interaction

- Web Content Accessibility Guidelines (WCAG) [8]—guidelines intended to improve information on a web site, including text, images, videos, etc.

- Accessible Rich Internet Applications (WAI-ARIA) [16]—defines a way to make dynamic web content and web applications based on new interactive technologies as Ajax, HTML5 more accessible

2.2. Accessibility Requirements For Content And User Interfaces

Learning management systems (LMS) work with web technology, so their user interfaces can be evaluated based on the basic principles for creating accessible web content as presented in WCAG 2.0. The universality of these guidelines is evidenced by the fact that it was approved in 2012 as an international standard: ISO/IEC DIS 40500 [8]. WCAG 2.0 identifies twelve guidelines and numerous compliance criteria (*"success criteria"*). WCAG 2.0 is based around four main principles, which provide the necessary basis for anyone to access and use a system. The four principles are described as follows:

- Perceivable: This principle is related to how information and user interface components must be presentable to users in ways they can perceive without limitations. This means that users must be able to perceive the content and information available in a web, the information presented in any part of the web must be visible to all of their senses.

- Operable: This principle is based on the fact that user interface components and navigation through a web must be operable. This is important so that users must be able to operate the interface, avoiding to ask the user some interaction that she cannot perform.

- Understandable: This means that users must be able to understand the information as well as the operation of the user interface without more details provided.

- Robust: Content presented in a web must be really robust, in a way that it can be interpreted easily by a wide variety of user agents, especially software and hardware prepared as assistive technologies. This means in other works that users must be able to access the content independently as technologies advance and evolve.

Under each of the four principles, there is a list of guidelines that address the principle. There are a total of 12 guidelines. One of the key objectives of the guidelines is to ensure that content is directly accessible to as many people as possible. There are success criteria related to each guideline, which describe specifically what must be achieved in order to conform to the WCAG 2.0 standard [8]. Each success criterion is written as a statement that will be either true or false when specific web content is tested against it. Table 1 presents the 12-guideline part of the standard.

Table 1. Accessibility Guidelines For Web Content Wcag 2.0

Principles	Guidelines
Perceivable	1.1 Provide text alternatives for any non-text content so that it can be changed into other forms people need, such as large print, Braille, speech or simpler language.
	1.2 Provide alternatives for time-based media.
	1.3 Create content that can be presented in different ways (for example, simpler layout) without losing information or structure.
	1.4 Make it easier for users to see and hear content, including separating foreground from background.
Operable	2.1 Make all functionality available from a keyboard.
	2.2 Provide users enough time to read and use content.
	2.3 Do not design content in a way that is known to cause seizures.
	2.4 Provide ways to help users navigate, find content, and determine where they are.
Understandable	3.1 Make text content readable and understandable.
	3.2 Make web pages appear and operate in predictable ways.
	3.3 Help users avoid and correct mistakes.
Robust	4.1 Maximize compatibility with current and future user agents, including assistive technologies.

The group of principles, guidelines, and success criteria based on WCAG 2.0 [8] are applicable to any web pages and digital content. In the case of e-learning systems (e.g., LMS), these systems are a group of web pages and educational digital content so WCAG 2.0 can be applied to each element. As a summary, the following six basic accessibility principles should be included in every e-learning system [17]:

1. Allow users to customize their portal based on their preferences.

2. Provide equivalents to every time-based media and visual elements.

3. Use different ways to present information in an interface.

4. Provide information appropriate compatible with assistive technologies.

5. Allow access to all functionalities via keyboard.

6. Provide background information and status and location information to the user at all times.

From WCAG 2.0 [8] guidelines and different accessibility related laws, in terms of basic functionality, e-learning systems (learning content management systems) should have the following basic characteristics:

1) Structure

   a) Absence of markup code errors in pages (HTML, CSS)

   b) Setting of accessibility preferences as default configuration, available for user personalization

   c) Accessibility check for content creators (HTML editors) and images selectors (e.g., alternative texts for each image)

   d) Summary of last activity within the system

2) Keyboard navigation

   a) Definition of a logical order to display tab indicators, provide a visual place mark to identify where the user is in a particular moment

   b) Provide links to jump to main content

   c) Functionality to simplify configuration to minimize secondary content pages and menus

   d) Functionality to select options using a simple combination of keys

   e) Provide complete access to all functionality via keyword, including HTML editors, controls in multimedia viewers, and Web 2.0 functionalities (e.g., Drag and drop")

   f) Enable keyboard shortcuts (hotkeys) and provide a definition page with all combinations

   g) Provide a complete sitemap structure for navigation in all systems

   h) If a key is pressed by mistake, provide the ability to undo and return to previous state

3) Magnification of screen size and functionality to change colors contrast

   a) Provide a standard design of the interface through all systems in order to find similar functionality on all tools

   b) Provide integration for assistive technologies

   c) Provide a selector to change style sheets for user personalization

   d) Avoid the communication of system information based on colors (e.g., buttons with a specific color and meaning)

   e) Provide the ability to change user preferences to change font size and style

   f) Maximize compatibility with assistive technologies

   g) Compliance-oriented design to improve interoperability with assistive technologies

   h) Consistent and unique design of headings, links, buttons and images description

   i) Provide descriptive forms including support for errors correction. Identification of the location of the users when filling a form

   j) Minimal use of frames, appropriate use of title in frames, provide adoption of ARIA standard attributes and navigational marks ("role landmarks"), structural tags, and alerts

4) Multimedia (audio) functionalities

2.3. Accessibility Requirements For Content Authoring Tools

Authoring tools are software and services included in e-learning systems (LMS), used for teachers and students to produce web content as educational material. Authoring tools related to LMS include desktop applications, multimedia authoring tools, and mainly HTML editors (e.g., what-you-see-is-what-you-get WYSIWIG editors). These tools should follow the Authoring Tool Accessibility Guidelines (ATAG) 2.0 [14].

The Authoring Tool Accessibility Guidelines (ATAG) explain to developers how to make and adapt the authoring tools to be accessible so that people with disabilities can access and create educational content. The guidelines explain how to help authors (teachers and students) to create more accessible web content (learning material) with inline validators, forms with hints and reminders.

Accessibility, from the perspective of authoring tools, is related to content creators and then for final users (especially people with disabilities). Thus, ATAG [14] is divided into two parts, each reflecting a key aspect of accessibility with respect to authoring tools. Part A "Make the authoring tool user interface accessible" relates to the accessibility of authoring tool user interfaces to authors with disabilities. Part B "Support the production of accessible content" relates to support by authoring tools for the creation, by any author (teachers and students, not just those with disabilities), of web content that is more accessible to end users with disabilities.

Besides general authoring tools, which are referred by ATAG, it is important to keep in mind that in the field of e-Learning, educational resources are usually packaged in containers for interoperability and reusability. Following ATAG [14] recommendations, tools used to prepare educational containers should take into account the accessibility requirements.

The format most commonly used is Sharable Content Object Reference Model (SCORM). This is a set of standards and specifications for creating structured teaching objects [18]. With SCORM, it is possible to create content that can be imported into different learning management systems providing SCORM compatibility. Based on the original definition of SCORM (ADL) [18], it is important to mention the six motivations of the standards: accessibility, adaptability, affordability, durability, interoperability, and reusability. In this chapter, Section 4 will elaborate on two aspects: accessibility and adaptability for the learning objects, building blocks for this standard.

2.4. Accessibility Requirements For Multimedia Tools

The users of an e-learning campus use different tools as media players, web browsers, and assistive technologies to be part of the educational process. These tools are known as user agents. The User Agent Accessibility Guidelines (UAAG) [15] explain how to make user agents accessible to people with disabilities, particularly to increase accessibility to web content, a basic building block for educational material in a virtual campus. As described in the working draft of UAAG Guidelines, in addition to helping developers of browsers and media players, UAAG 2.0 benefits developers of assistive technologies because it explains what types of information and control an assistive technology may expect from a user agent that follows UAAG 2.0. Assistive technologies not addressed directly by UAAG 2.0 [15] (e.g., Braille rendering) are still essential to ensuring web access for some users with disabilities.

UAAG is organized in guidelines, principles, and success criteria elements. There are five principles: "perceivable, operable, understandable, programmatic access, and specification and conventions." Following the principles, there are 27 guidelines [15].

2.5. Accessibility Requirements Of Dynamic Content And Rich User Interfaces

Nowadays, web applications, in our work the case of virtual campuses based on learning management systems, are increasingly using more advanced and complex user interface controls such as tree controls for site navigation, drag-and-drop functionality, or technologies developed with Ajax or DHTML. To prevent accessibility issues, the Web Accessibility Initiative (WAI) [13] proposed a recommendation called "Accessible Rich Internet Applications," usually known as WAI-ARIA [16]. This suite of recommendations defines a way to make web content and web applications more accessible to people with disabilities. It especially helps with dynamic content and advanced user interface controls developed with Ajax, HTML, JavaScript, and related technologies.

More specifically, WAI-ARIA provides a framework for adding attributes to identify features for user interaction, giving hints on how they relate to each other, and their current state. The WAI-ARIA framework [16] identifies innovative navigation techniques to mark regions and common web structures as menus, primary content, secondary content, banner information, and other types of web structures. As a working example for developers, with WAI-ARIA, it is possible to identify regions of pages and enable keyboard users to easily move among regions rather than having to press the tab key many times.

WAI-ARIA also includes technologies to map controls, Ajax live regions, and events to accessibility application programming interfaces (APIs), including custom controls used for rich Internet applications. WAI-ARIA [16] techniques apply to widgets such as buttons, drop-down lists, calendar functions, tree controls (for example, expandable menus), and others usually available in virtual campuses so it is important that LMS administrators.

3. KNOWLEDGE REQUIRED FOR USERS RELATED TO AN ACCESSIBLE VIRTUAL CAMPUS

Once a virtual campus reaches an acceptable level of accessibility, this accessibility must be constantly maintained. The content and learning material published by the teachers and administrators will be periodically updated, and it is important to teach stakeholders on how to create and adapt learning content to be accessible following most used guidelines. Among the actions to be carried out periodically to maintain accessibility in a virtual campus are the following:

- Training for teachers and students in techniques for creating accessible digital contents

- Training for teachers in Universal Learning Design techniques

- Providing in the virtual campus the functionality of online accessibility checkers when final users work with basic actions such as uploading images and alternative text, providing context information for links, validating information in content editors, etc.

3.1. Techniques for Creating Accessible Documents

It is important to take into consideration that when digital content is created by teachers or students in any type of format (textual, graphic, audio, or multimedia), it is necessary to keep in mind that final users of such content may be people with physical, sensory, or cognitive limitations, who could find barriers to access the information. In fact, at some point in our lives, we all probably will have limitations that can affect our access to digital content. Among the difficulties that teachers are facing when preparing learning content in digital format is the diversity of authoring tools available to create the content. In [19], a collection of the basic considerations to create accessible digital content are presented and for diversity, the Accessible Digital Office Document (ADOD) initiative [20] prepared different recommendations based on the content creator used.

The Accessible Digital Office Documents (ADOD) Project [20] is an initiative created to provide guidelines on the accessibility of office documents, office document formats, and office applications independent of the tool used to create the content. ADOD provides both an "ADOD Assessment Framework" and a suite of practical guidance documents that are intended to help stakeholders in the educational process to make decisions about office applications. Currently, ADOD is based primarily on the WCAG and ATAG recommendations presented in Section 2.

The recommendations provided for office tools are also applicable to PDF documents. Among the recommendations to create accessible PDF documents with learning content, based on WCAG 2.0 guidelines [21], are the following:

1. Check that all nontext elements should include alternative text.

2. Check for background color and foreground contrast.

3. Specify the text language in all documents to help assistive technologies.

4. Check if hyperlinks are correctly formatted and functional.

5. Provide labeling of elements and correct use of styles.

6. Provide alternative texts and contextual information for hyperlinks.

7. Provide information for abbreviations and acronyms.

8. Check for language changes in the text if more than one language is used.

9. Identify decorative elements in headers and footers.

10. Add markers (bookmarks) that allow the user to jump to a specific part of the document.

11. Verify that the default reading order, according to the structure of tags, makes sense and is consistent.

12. Check for the proper security settings, avoiding sharing a document with password.

13. If the PDF contains an image from a scanned document, an OCR process has to be prepared to provide the text as background alternative for assistive technologies.

14. In case the PDF contains a form, the fields properties should have a detailed description to help the user to fill in the requested information.

Besides the ADOD project and the recent book [19], other initiatives and guidelines for creating electronic documents accessible are found in [22–24].

As an alternative, authors can export a document in DAISY format, which is a good way to ensure that a document is accessible. DAISY is a multimedia format that maintains and promotes a system of Access to standard printed documents for blind, low vision or other problems. The format was developed by the DAISY consortium in 1996 and is currently based on the definition of ANSI/NISO Z39.86-2005 standard [25].

The text content can be exported in DAISY format with plug-ins for Word processors as Microsoft Office Word and LibreOffice Writer. This format can be tested with a DAISY complaint software, for example, the AMIS software (http:/www.daisy.org/amis). Exporting content to DAISY [25] format allows authors to check the accessibility of a document to a person with vision problems because the software prepares and audio book based on the content.

Administrators for a virtual campus based on learning management systems (LMS) should not assume that the users (e.g., teacher, instructor, tutor, student, etc.) have all the knowledge concerning WCAG guidelines or principles of Universal Learning Design. It is important to incorporate and provide descriptive aid in the different interfaces and provide validators to allow users to know whether the content is accessible based on the minimal requirements established by the educational institution.

Examples of basic functionality to be included to help final users creating contents are as follows:

- Basic code validator (HTML) included in WYSIWYG content editors usually used in application for discussion forums, wikis, information box, etc. (e.g., AChecker plug-in (www.achecker.ca) for ATutor LMS)

- Validator for images and alternative text aids for users editing content

- Validator for accessibility in equation writer editors

3.2. Automatic Analysis Using Validation Tools

The evaluation of the accessibility of a virtual campus and its contents is performed in two main phases.

1. Automatic analysis with validation tools

2. Manual analysis/heuristic evaluation by experts and end users

The first phase is proposed to use an online automatic validator based on the WCAG guidelines. Some of the identified tools available online are as follows:

- Examinator (based on WCAG 2.0 guidelines) (www.examinator.ws)

- AChecker (based on WCAG 2.0, HTML y CSS) (www.achecker.ca)

- TAW (based on WCAG 2.0) (www.tawdis.net)

- CynthiaSays (based on WCAG 2.0) (www.cynthia-says.com)

- Tingtun (based on WCAG 2.0) (accessibility.tingtun.no)

- HERA (based on WCAG 1.0) (www.sidar.com/hera)

- WebAim (Web Accessibility Evaluation Tool) (http://wave.webaim.org)

- HTML validator (http://validator.w3.org/)

- CSS validator (http://jigsaw.w3.org/css-validator/)

The assessment of accessibility should identify a simple of pages related to the main actions from users within the virtual campus. The main actions to be evaluated are as follows:

1) Start using the virtual campus.

   a) Visit the homepage of the educational institution.

   b) Visit the accessibility information for the educational institution.

   c) Pages that the user needs to visit to reach the virtual campus login pages.

   d) Registration, enrollment, and log into the virtual campus.

   e) Change the personal settings and preferences for the user.

   f) Follow the steps to visit a course page.

2) Use basic functionality for students.

   a) Find and review content within a course, including multimedia content.

   b) Contribute to course content assigned to the student (wiki tool or upload a file form).

   c) Find, check, and submit and assessment.

   d) Find a questionnaire, read the instructions, answer all questions, and send the completed questionnaire (quiz).

   e) Find and check the gradebook.

   f) Read news and announcements published by the teacher.

g) Find, publish, and interact in a course blog.

h) Find the discussion forums application and be part of a conversation.

3) Use basic functionality for teachers.

a) Create and publish content in a course page.

b) Create content on the course with conditional availability (hide and enable content).

c) Create a task assignment.

d) Create a questionnaire with different types of questions.

e) Reorganize and sort items in the course menu.

f) Copy items from one section of the course to another section.

g) Login and manage the student gradebook.

h) Evaluate and comment a student assignment.

The pages included in the virtual campus (dynamic content and login required pages) usually cannot be verified easily by automatic analysis tools. To perform this analysis, it is possible to use installed tools as plug-ins (e.g., WAVE tool) or download the pages to be evaluated as static content.

The second phase of the evaluation is the heuristic evaluation by experts and end users. Automatic validation tools offer a partial view on the accessibility, but it is important to have a group of accessibility experts and final users with disabilities to test the main functions and have a contrasted opinion and recommendations to improve the accessibility of the virtual campus.

4. REQUIREMENTS TO CREATE ACCESSIBLE LEARNING OBJECTS

Learning objects (LOs) are the minimum unit in which educational content is organized so that it can be easily published for a better understanding. One of the most popular definitions of LO is that offered by Wiley "as any digital resource that can be reused to support learning" [26].

The main goal of an LO is their reuse in more than one training activity. To do this, it is necessary that the LO can be found in a simple manner. To achieve this, we need to describe the LO's characteristics, including their metadata, which are a set of fields that provide information about the LO such as, for example, its title, its description, the language in which it is written, or its scope. There are some specifications and standards commonly used to define the LO metadata for their correct description. The most popular are Dublin Core [27] and LOM [28].

LOs, besides regular metadata, must have associated accessibility metadata that describe their accessibility characteristics and that make them accessible to all people. These metadata are the fields used for searching accessible LOs.

Repositories are used to store LOs and to facilitate their search and therefore their reuse. Search operations are performed based on their metadata, hence the importance of clearly and correctly describing the resources, which provides more precise searches. One of the most known repositories is Merlot [29], which have an interesting advanced search function.

When users need to perform a training activity, they use these repositories to find the learning objects that better adapt to that training, thus drawing up a new course from the learning objects found in the repository or repositories to they can access.

Metadata should be inserted in an XML (extensible markup language) file [30], composed of each of the fields (each field corresponds to a metadata) described following one of the standards published for this purpose, such as, for example, learning object metadata (LOM) [28]. This work is provided by metadata editors such as, for example, LomPad, known for being one of the most used [31].

As shown in Figure 1, the LomPad editor allows completing the LOM metadata fields. Once all data have been inserted, an XML file containing all information is generated.

The process for sharing content and distributing it among different information systems is to pack it in a compressed file composed of the content and metadata that describe it. In this scope, there are two specifications widely used, such as Sharable Content Object Reference Model (SCORM) [18] and IMS Common Cartridge [32]. Just as there are editors to help content authors to describe the metadata, there are also editors that help to pack this content along with metadata. One of the most known editors is Reload Editor [33].

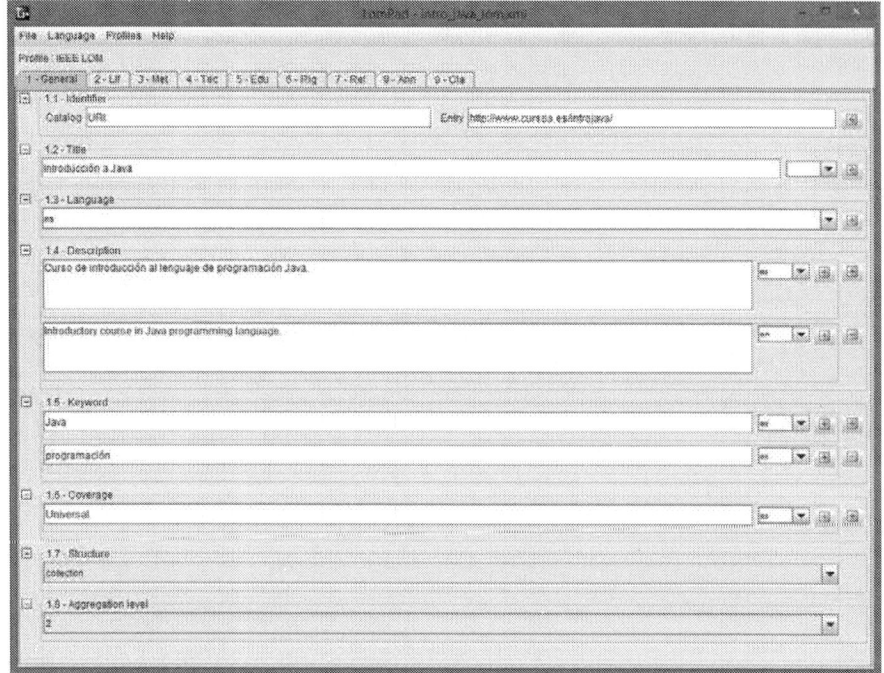

Figure 1. LomPad editor.

Reload not only allows packing content based on SCORM specification but also allows to describe resources with metadata (analogously to LomPad) and to organize the sequencing of these resources.

4.1. Ims Access For All (Afa) V3.0

IMS AfA v3.0 specification [12] is a way to add accessible metadata to a learning object. Using this, we can describe what is the best sensory form to access the learning object. The specification is created with the aim of simplifying the ISO/IEC 24751 standard [9–11] due to the difficulties encountered when putting it into practice. Both standard and specification in version 3.0 cover the entire process from reading the user needs to the search mechanism needed to find the LO that meets those needs or preferences. The main objectives of IMS AfA v3.0 specification are as follows [12]:

- Being simple and easy to understand
- Facilitating its modification to suit the needs of the organizations requiring some parts of the model
- Facilitating integration with other metadata and specifications
- Allowing integration with devices' properties standards for accessibility
- Allowing integration with user agents, accessibility APIs, and productivity-oriented accessibility standards
- Allowing inclusion in accessibility frameworks and tools

It has two metadata models to describe the following:

- Personal needs and preferences (PNP): description model of the users' needs and preferences to access and interact with the digital resources
- Digital resource description (DRD): description model of the accessibility metadata for the digital training resources

With the AfA DRD, the accessible metadata of the learning objects are described and with the AfA PNP the students can provide their personal needs (or those due to disability environments). The goal is to find the learning objects that best match user needs and preferences in an automated way, solving the metadata similarities between PNP and DRD.

4.1.1. Digital Resource Description (Drd)

AfA DRD defines the accessibility metadata of a resource that will be used for searching and using the most adequate learning resource to each user according to his or her PNPs.

The adaptation of a learning object occurs when we produce one with the same training content but with a different form of access. To achieve this, two types of LOs must exist: original and adapted. An original resource corresponds to a primary resource, while an adapted resource presents the same educational information than the original resource, for example, a PDF format file as the original resource and an audio description of its content as an adapted resource. The first one presents textual access, while the adaptation presents auditory access to the same educational content.

Original resources may have any number of adaptations, which may be total or partial, i.e., or they are adaptations of the whole educational content or they are just a part of this.

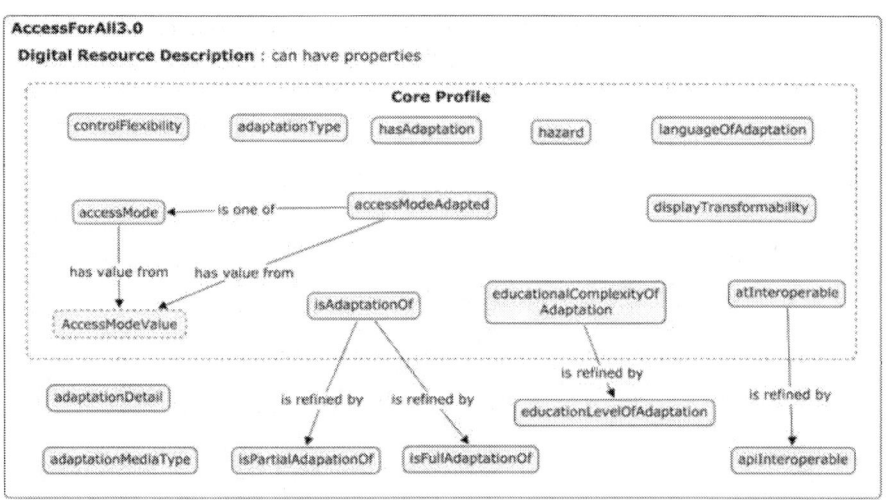

Figure 2. Digital Resource Description (Drd) Properties.

Figure 2 shows the accessibility properties or metadata of a resource and how they relate to each other, as IMS AfA v3.0 specification presents them. As seen in the figure, in order to simplify as much as possible the data model, the metadata have been organized in two clearly distinguished levels:

1. Those belonging to a basic core (Core Profile), containing the most important metadata, necessary for a proper description of the resource

2. Those belonging to the full specification, which extent and complement the basic core information

4.1.2. Personal Needs And Preferences (Pnp)

The specification shows a common information model to define and describe the student's or user's PNPs with a different sensory perception mode or who is in a disability context. The user's PNPs may be environmental (for example, "in the dark"), may be related to the communications technology or the available and specific information services (for example, "when a Braille device is available"), or may relate to social situations (for example, "when my nurse is present") or other scenarios.

The recommended method to generate the student's PNPs is the presentation of a form with various options (like aforementioned or preferred sensory mode). The PNPs will be generated from students' responses to these questions.

The declaration of PNPs is associated to one person. In turn, one person can generate several sets of PNPs for being used in the environment he or she is at each moment (for example, in the dark or in a noisy area). Like any software application, user's PNPs should be easily modified by editing the user profile and by allowing its extension, replacement, or removal.

Figure 3 shows the user's accessibility properties and how they relate to each other. In the same manner as specification AfA DRD, there are properties belonging to the basic profile (Core Profile) and those belonging to the full specification.

4.2. Application Scenario

In this section, a scenario of use of IMS AfA v3.0 specification [12] is described in addition to other e-learning specifications and standards previously explained, describing all stages for getting an accessible learning object.

First, a content author plans to carry out a learning resource that contains a video tutorial (original resource) of an educational course. An alternative content (adapted resource) is created to provide access to this resource to the students with disabilities (especially those with visual problems). This resource consists of an audio description (audio file that describes the images containing meaningful information).

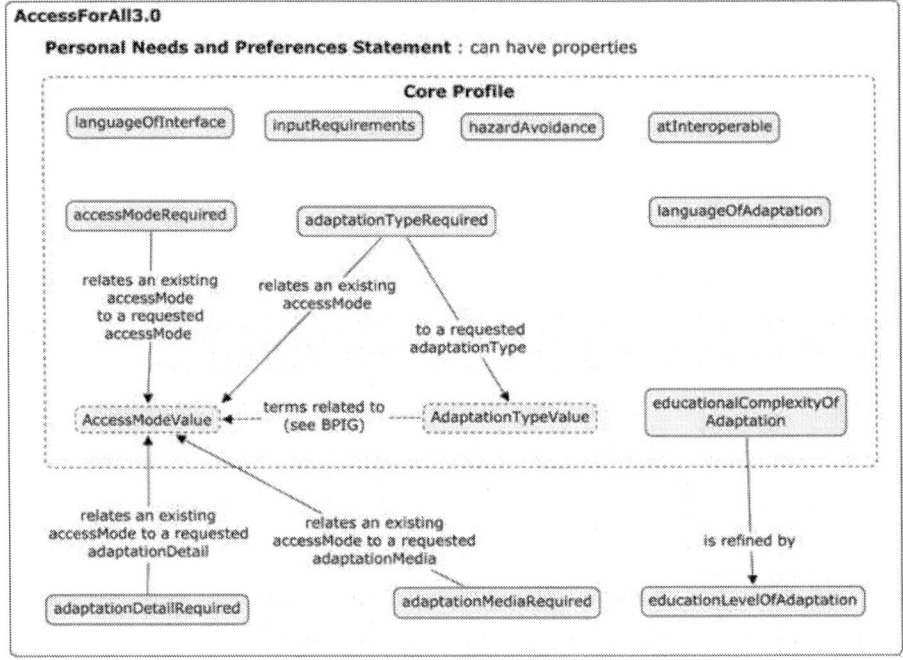

Figure 3. Personal Needs And Preferences (Pnp) Properties.

The content author uses LomPad [31] or Reload [33] to describe the LOM metadata of the video tutorial, thus describing the educational material so that it can be located and reused in different training activities.

Then it is necessary to include the accessibility metadata of the original resource; thus, the type of sensorial perception is described, which is needed to understand the training content. As this is a video, both the visual and the auditory senses are needed. For inserting the accessibility metadata by following IMS AfA specification, the author can use LomPad-AfA tool [34], as shown in Figure 4, whose ultimate goal is to generate the XML file, as shown in Figure 5. LomPad-AFA allows the content authors and the learning platform users to insert accessibility metadata of LOs (DRDs) and students' PNPs, respectively, generating both XML format files. This tool allows to complete the properties of DRDs and PNPs graphically and to generate the corresponding XML file following the IMS AfA v3.0 specification.

In the XML file generated, which is shown in Figure 5, it is described that the original resource has two access modes: visual and auditory. It has one adaptation: OR_1_A1, and it can be controlled using the keyboard and mouse.

The following step will be creating the description for the adapted resource, which contains the audio description. Using LomPad-AfA, the accessibility metadata are filled and the XML file is generated (Figure 6).

Figure 4. Original Resource's Drd Xml (Afadrdv3p0_Or_1.Xml).

In the XML file generated, which is shown in Figure 6, it is described that the adapted resource has an auditory access mode, and it adapts a visual one. More details about the type of adaptation are given through property "adaptation type," and it is specified that it is an audio description. It has full control by keyboard and mouse. It is an adaptation of the original resource OR_1, and it is a partial adaptation. Finally, it states that the audio is recorded using a human voice.

Once the resources are created and the metadata are defined in their corresponding XML files, a package containing all information and following SCORM specification will be created. As shown inFigure 7, the SCORM package will be composed of two resources (the original and the adaptation) and their metadata files. The original resource will have associated two metadata files, one with its LOM metadata and another one with the IMS AfA metadata. Adapted resource only needs the IMS AfA metadata since the adapted resource contains the same learning information as the original.

```
<?xml version="1.0" encoding="UTF-8"?>
<accessForAllResource xsi:schemaLocation="http://www.imsglobal.org/xsd/accessibility/afadrdv3p0/imsafav3p0drd_v1p0
imsafav3p0drd_v1p0.xsd" xmlns:xsi="http://www.w3.org/2001/XMLSchema-instance"
xmlns="http://www.imsglobal.org/xsd/accessibility/afadrdv3p0/imsafav3p0drd_v1p0">
    <accessMode>afaterms-visual</accessMode>
    <accessMode>afaterms-auditory</accessMode>
    <hasAdaptation>OR_1_A1</hasAdaptation>
    <controlFlexibility>afaterms-fullKeyboardControl</controlFlexibility>
    <controlFlexibility>afaterms-fullMouseControl</controlFlexibility>
</accessForAllResource>
```

Figure 5. Adapted Resource's A1 Drd Xml (Afadrdv3p0_Or_1_A1.Xml).

```
<?xml version="1.0" encoding="UTF-8"?>
<accessForAllResource xsi:schemaLocation="http://www.imsglobal.org/xsd/accessibility/afadrdv3p0/imsafav3p0drd_v1p0 imsafav3p0drd_v1p0.xsd"
xmlns:xsi="http://www.w3.org/2001/XMLSchema-instance" xmlns="http://www.imsglobal.org/xsd/accessibility/afadrdv3p0/imsafav3p0drd_v1p0">
    <accessMode>afaterms-auditory</accessMode>
    <accessModeAdapted>afaterms-visual</accessModeAdapted>
    <adaptationType>afaterms-audioDescription</adaptationType>
    <languageOfAdaptation>en</languageOfAdaptation>
    <controlFlexibility>afaterms-fullKeyboardControl</controlFlexibility>
    <controlFlexibility>afaterms-fullMouseControl</controlFlexibility>
    <isAdaptationOf>OR_1</isAdaptationOf>
    <isPartialAdaptationOf>OR_1</isPartialAdaptationOf>
    <adaptationDetail>afaterms-recorded</adaptationDetail>
</accessForAllResource>
```

Figure 6. Lompad-Afa Resource Drd Properties.

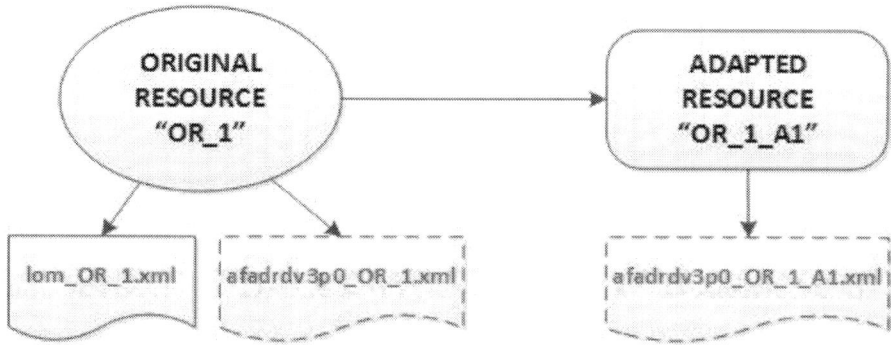

Figure 7. Lompad-Afa User Pnp Properties.

Furthermore, LomPad-AfA tool allows generating XML files containing the users' PNPs. For example, if a blind person or a person with visual problem wants to describe his or her preferences, he or she has to fill the metadata, as shown in Figure 8, and generate the XML file, as shown in Figure 9.

In the XML file generated, as shown in Figure 9, it is described that, for visual content, the user prefers adapted resources that have an auditory or textual access mode. By means of property "adaptation type required," more details about the type of desired adaptation for visual content are given, and it is specified that they should contain audio description or long description. A learning system (educational platform, learning object repository, etc.) that is able to understand the PNP defined above and whose user is interested in learning the educational resource of the video tutorial, which represents the original resource, should show the adaptations that are associated with it.

5. CONCLUSIONS

The accessibility of a virtual campus should be ensured at two levels: (1) the accessibility of the learning management system (LMS) that supports the campus and (2) the accessibility of the learning materials published on the platform. A virtual campus with an LMS platform that meets the criteria under different guidelines as described in WCAG 2.0 will be accessible, but when new content is published, the accessibility could be lost, and students with disabilities could face barriers to achieve the learning objectives. Thus, it is important to maintain a continuous process of training for stakeholders involved in the virtual campus.

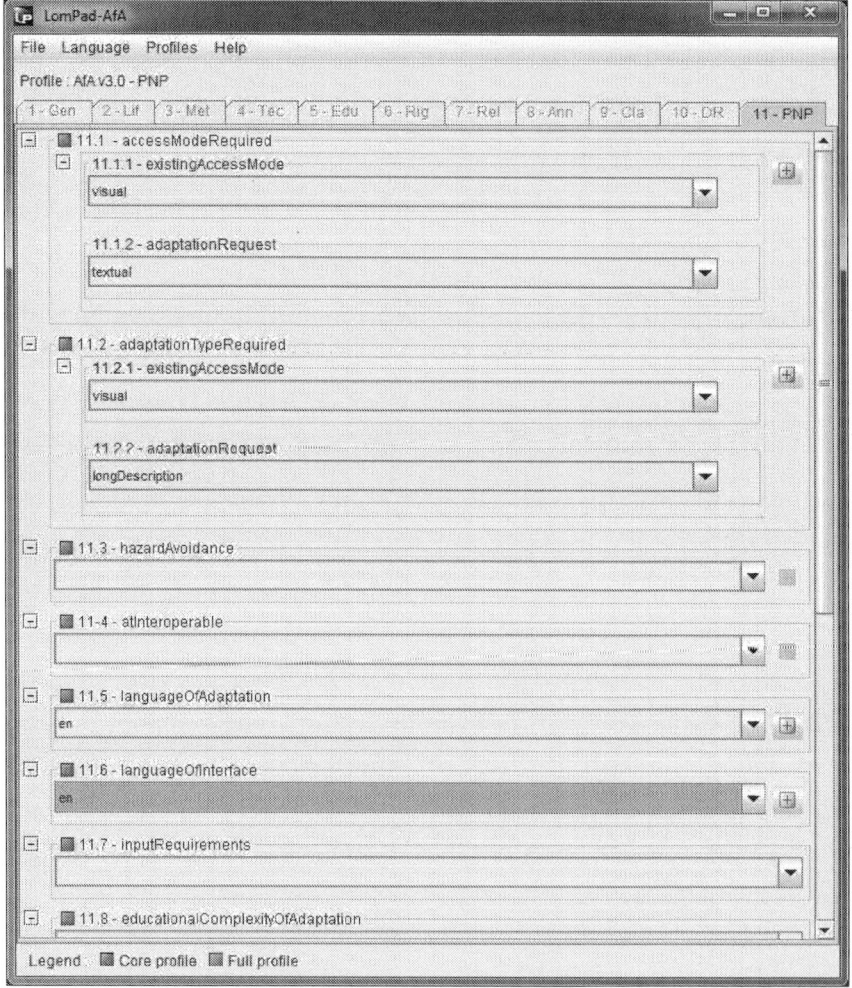

Figure 8. Scorm Content.

The main principles that an accessible virtual campus should provide are as follows:

(1) allow users to customize their portal based on their preferences,

(2) provide equivalents to every time-based media and visual elements,

(3) use different ways to present information in an interface,

(4) provide information appropriate compatible with assistive technologies,

(5) allow access to all functionalities via keyboard, and

(6) provide background information and status and location information to the user at all times.

```xml
<?xml version="1.0" encoding="UTF-8"?>
<accessForAllUser xsi:schemaLocation="http://www.imsglobal.org/xsd/accessibility/afapnpv3p0/imsafa3p0pnp_v1p0
imsafa3p0pnp_v1p0.xsd" xmlns:xsi="http://www.w3.org/2001/XMLSchema-instance"
xmlns="http://www.imsglobal.org/xsd/accessibility/afapnpv3p0/imsafa3p0pnp_v1p0">
    <accessModeRequired>
        <existingAccessMode>afaterms-visual</existingAccessMode>
        <adaptationRequest>afaterms-textual</adaptationRequest>
    </accessModeRequired>
    <accessModeRequired>
        <existingAccessMode>afaterms-visual</existingAccessMode>
        <adaptationRequest>afaterms-auditory</adaptationRequest>
    </accessModeRequired>
    <adaptationTypeRequired>
        <existingAccessMode>afaterms-visual</existingAccessMode>
        <adaptationRequest>afaterms-audioDescription</adaptationRequest>
    </adaptationTypeRequired>
    <adaptationTypeRequired>
        <existingAccessMode>afaterms-visual</existingAccessMode>
        <adaptationRequest>afaterms-longDescription</adaptationRequest>
    </adaptationTypeRequired>
    <languageOfAdaptation>en</languageOfAdaptation>
    <languageOfInterface>en</languageOfInterface>
    <adaptationDetailRequired>
        <existingAccessMode>afaterms-visual</existingAccessMode>
        <adaptationRequest>afaterms-enhanced</adaptationRequest>
    </adaptationDetailRequired>
</accessForAllUser>
```

Figure 9. User Pnp Xml (Afapnpv3p0_Usr1.Xml).

Training for users of virtual campus, publishing learning content is an ongoing process that should primarily include the following components:

(1) training teachers and students in techniques for creating accessible documents,

(2) training teachers on universal learning design techniques, and

(3) training LMS administrators to maintain the accessibility and configure the LMS to provide validators of accessibility in content editors, to ease the process of learning objects publication.

IMS AfA v3.0 specification presents to the content authors and developers the technical way to follow for achieving an accessible online teaching. According to ISO/IEC 24751-2-3 standard and IMS AfA v3.0 specification, the basic steps in developing an accessible online course are as follows:

creating accessible learning objects (LOs), both original and adapted, by means of inserting the accessibility metadata; reading the users' personal needs and preferences (PNP); and searching and presentation of LOs meeting those PNPs.

For an LO that can be used in an educational platform, it is necessary to pack all files shaping the LO with the files containing its metadata, including the accessibility ones, and following the standards established. There is a great lack of technical applications and human resources to provide assistance in developing accessible resources.

6. ACKNOWLEDGEMENTS

This work is funded by the University of Alcalá (Spain) and Galileo University (Guatemala) (grant ESVIAL project). Authors also want to acknowledge support from the Master in Software Engineering for the Web and the TIFyC research group.

REFERENCES

1. Power C., Petrie H., Sakharov V., Swallow D.. Virtual learning environments: another barrier to blended and e-learning.. In: Miesenberger K., Klaus J. Zagler W., Karshmer A.I., editors. ICCHP2010; Springer; 2010.
2. World Wide Web Consortium. How People with Disabilities Use the Web [Internet]. [Updated: 2014]. Available from: http://www.w3.org/WAI/intro/people-use-web/ [Accessed: 01-03-2015]
3. Amado-Salvatierra H. R., Hernández R., García-Cabot A., García-López E., Batanero C., Otón S.. Towards a methodology for curriculum development within an accessible virtual campus. In: Computers Helping People with Special Needs; 2014; Springer; 2014. pp. 338–341.
4. Armstrong H.L.. Advanced IT education for the vision impaired via e-learning. Journal of Information Technology Education. 2009;8(1):243–256.

5. Burgstahler S., Corrigan B., McCarter J.. Making distance learning courses accessible to students and instructors with disabilities: a case study. Internet and Higher Education. 2004;7(3):233–246.

6. Edmonds C.D.. Providing access to students with disabilities in online distance education: legal and technical concerns for higher education.. American Journal of Distance Education. 2004;18(1):51–62.

7. United Nations. Convention on the Rights of Persons with Disabilities [Internet]. 2008. Available from: http://www.un.org/disabilities/default.asp?id=150 [Accessed: 01-03-2015]

8. ISO. ISO/IEC 40500:2012. W3C Web Content Accessibility Guidelines 2.0 [Internet]. 2012. Available from: http://www.iso.org/iso_catalogue/catalogue_tc/catalogue_detail.htm?csnumber=58625 [Accessed: 01-03-2015]

9. International Standard Organization. ISO/IEC 24751-1:2008, Information Technology—Individualized Adaptability and Accessibility in E-learning, Education and Training. Part 1: Framework and Reference Model. [Internet]. [Updated: 2008]. Available from: http://www.iso.org/iso/catalogue_detail?csnumber=41521 [Accessed: 01-03-2015]

10. International Standard Organization. ISO/IEC 24751-2:2008, Information Technology—Individualized Adaptability and Accessibility in E-learning, Education and Training. Part 2: "Access for All" Personal Needs and Preferences for Digital. [Internet]. [Updated: 2008]. Available from: http://www.iso.org/iso/catalogue_detail?csnumber=41521 [Accessed: 01-03-2015]

11. International Standard Organization. ISO/IEC 24751-3:2008, Information Technology—Individualized Adaptability and Accessibility in E-learning, Education and Training. Part 3: "Access for All" Digital Resource Description. [Internet]. [Updated: 2008]. Available from: http://www.iso.org/iso/catalogue_detail?csnumber=41521 [Accessed: 01-03-2015]

12. IMS Global Learning Consortium, Inc.. IMS Access For All Version 3.0 [Internet]. Available from: http://imsglobal.org/accessibility [Accessed: 01-03-2015]

13. World Wide Web Consortium. WAI Guidelines and Techniques [Internet]. [Updated: 2014]. Available from: http://www.w3.org/WAI/guid-tech.html [Accessed: 01-03-2015]

14. World Wide Web Consortium. Authoring Tool Accessibility Guidelines (ATAG) 2.0 [Internet]. [Updated: 2014]. Available from: http://www.w3.org/TR/ATAG20/ [Accessed: 01-03-2015]

15. World Wide Web Consortium. User Agent Accessibility Guidelines (UUAG) 2.0. Working Draft [Internet]. [Updated: 2014]. Available from: http://www.w3.org/TR/UAAG20/ [Accessed: 01-03-2015]

16. World Wide Web Consortium. Accessible Rich Internet Applications (WAI-ARIA) 1.0 [Internet]. [Updated: 2014]. Available from: http://www.w3.org/TR/wai-aria/ [Accessed: 01-03-2015]

17. Moodle Community. Especificación de accesibilidad Moodle [Internet]. [Updated: 2012]. Available from: http://docs.moodle.org/dev/Moodle_Accessibility_Specification [Accessed: 01-03-2015]

18. Advanced Distributed Learning (ADL). Sharable Content Object Reference Model (SCORM). http://www.adlnet.org/scorm/

19. Hilera J.R., Campo E., García-López E., García-Cabot A.. Guide for Creating Accessible Digital Content: Documents, Presentations, Videos, Audios, and Web Pages. Alcalá de Henares, España: Universidad de Alcalá; 2015. DOI: 978-84-15834-82-3

20. Inclusive Design Research Centre. Accessible Digital Office Document (ADOD) Project [Internet]. [Updated: 2010]. Available from: http://adod.idrc.ocad.ca/ [Accessed: 10-02-2015]

21. World Wide Web Consortium. Técnicas PDF accesibilidad WCAG 2.0 [Internet]. [Updated: 2014]. Available from: http://www.w3.org/WAI/GL/WCAG20-TECHS/ [Accessed: 01-03-2015]

22. Australian Agency for International Development.. Guidelines for preparing accessible content [Internet]. [Updated: 2013]. Available from: http://www.ausaid.gov.au/business/Pages/web-content-accessibility-guidelines.aspx [Accessed: 01-02-2015]

23. Moreno L.. Recursos para elaborar documentación accesible.. Universidad Carlos III, España; 2013.

24. Sama V., Sevillano E.. Guía de accesibilidad de documentos electrónicos. España: Universidad Nacional de Educación a Distancia, España.; 2012.

25. DAISY. DAISY Consortium—Creating the Best Way to Read and Publish [Internet]. [Updated: 2014]. Available from: http://www. daisy.org/ [Accessed: 01-02-2015]

26. Wiley D. Connecting Learning Objects to Instructional Design Theory: A Definition, a Metaphor, and a Taxonomy, in The Instructional Use of Learning Objects.. Reusability:2001.

27. Dublin Core Metadata Initiative. Dublin Core [Internet]. Available from: http://dublincore.org/ [Accessed: 01-03-2015]

28. Institute of Electrical and Electronics Engineers. IEEE Learning Object Metadata (LOM) [Internet]. [Updated: 2002]. Available from: http:// www.ieeeltsc.org/ [Accessed: 01-03-2015]

29. MERLOT cooperative. MERLOT (Multimedia Educational Resource for Learning and Online Teaching) [Internet]. Available from: http://www. merlot.org/ [Accessed: 01-03-2015]

30. World Wide Web Consortium. Extensible Markup Language (XML). [Internet]. [Updated: 2008]. Available from: http://www.w3.org/XML/ [Accessed: 01-10-2015]

31. Licef. LomPad, learning object metadata editor. [Internet]. [Updated: 2004]. Available from: http://helios.licef.ca:8080/LomPad/en/index.htm [Accessed: 01-05-2014]

32. IMS Global Learning Consortium, Inc.. IMS Common Cartridge [Internet]. Available from: http://www.imsglobal. org/commoncartridge.html [Accessed: 01-03-2015]

33. Liber O., Olivier B., Beauvoir P.. Reusable eLearning Object Authoring & Delivery [Internet]. [Updated: 2013]. Available from: http:// www.reload.ac.uk/ [Accessed: 01-02-2015]

34. Otón S., García A., García E., Barchino R., Amado-Salvatierra H.. Transforming LOMPad to Support IMS Access for All v3.0. In: Advanced Learning Technologies (ICALT), 2014 IEEE 14th International Conference on; 2014; IEEE; 2014. pp. 599–600.

CHAPTER 11

Development of an Asynchronous Web Based E-Learning System

Nathaniel Olufisayo Oluwaniyi, Babajide Olakunle Afeni, Olusegun Olayinka Lawal

Department of Computer Science, Joseph Ayo Babalola University (JABU), Ikeji-Arakeji, Nigeria

ABSTRACT

Advancements in Information Communication Technology (ICT) have led to several opportunities especially the ones provided by the Internet. Several people are now taking advantage of distance learning courses and in the past few years huge research efforts have been dedicated to the development of distance learning systems. So far, many e-learning systems are proposed and used practically. This paper focused on the development of an asynchronous and interactive Web-based e-learning system. Its primary objective is to develop a fast, reliable, effective and efficient web- based e-learning system that will address the problems associated with the traditional learning system. Succinctly, the paper discusses the design of a system that enhances e-learning where course lecturers can set their courses, tests and quizzes at their convenient time and can track the activities and performance of their students and guide them to acquire knowledge without being obliged to be physically present on the institution campus. The system was designed using PHP as the scripting language, Macromedia Dreamweaver for the web page, MySQL as the database and Apache as the web server. The system was implemented using real data and the result was successful. This system is no doubt a solution to the constraints of the classical learning system and can be used successfully in distance learning, training, and various educational institutions.

KEYWORDS

E-Learning, Web-Based, Asynchronous, Information Communication Technology (ICT)

1. INTRODUCTION

The Internet has profoundly changed the way we communicate and interact with one another. Studies conducted by Pew Internet and American Life Project found that as of June 2005, 137 million Americans (or 68% of American adults) used the Internet, up from 63% one year ago [1] . About 94 million Americans used the Internet for such daily activities as e-mailing, searching for information, getting news, checking the weather, instant messaging, and online banking, to name a few [2] . According to [3] , the Internet has brought dramatic changes to education. The educational use of computers has evolved from numerical calculations, spreadsheets, and word processors to multimedia and web-based applications. The introduction of the Intranet/Internet access in institutions of learning and organizations has encouraged the development of new mediums and systems within the scope of education and training. Due to the opportunities provided by the Internet, more and more people are taking advantage of distance learning courses [4] . As a result, in a new era, learning and training approaches have emerged, where students can learn independently at any time and from any location, simply by using their computers connected to the net along with the appropriate systems and tools. Teachers also can teach from on-line location and can schedule their lectures and examination without the classical physical constraints. This electronic learning approach known also as e-learning, has opened new horizons in teaching for both teachers and students. E-learning is a general term used to refer to a form of learning in which the instructor and student are separated by space or time where the gap between the two is bridged through the use of online technologies. An e-learning system is a powerful integrated system that supports a number of activities performed by teachers and students during the e-learning process. A web-based asynchronous e-learning system allows educators to develop web-based course notes, test and examination, to communicate with students and to monitor their progress. Students engage the system fort for learning, communication and collaboration. It is important to note that the effectiveness of an e-learner's experience is greatly enhanced through student-centered design. For example, students remember more information from a textbook that is well organized, with extensive visuals, reflection/ interaction points, clear headings, etc. The same concept exists for online courses-learners learn better through use of clear headings, limited distracters, visuals, screen-friendly fonts, appropriate white space, web safe colors, etc. This work developed, implemented and proposed a simple web-based asynchronous e-learning system, in which educators can set teach and also assess their students in form of quizzes and exams. The system will allow educators to store, update, and delete questions from the database using the web, in a very easy and simplified manner. Also, they will be able to track the activities of their students and can guide them to reach the pre-determined objectives of the courses.

1.1. Statement of Problem

Learning is commonly defined as a process that brings together cognitive, emotional, and environmental influences and experiences for acquiring, enhancing, or making changes in one's knowledge, skills, values, and world views. It is the process of gaining knowledge or skill by reading and studying, from experience, from being taught, etc. Learning system refers to the method or way of impacting knowledge. The two main types of learning system are identified in the educational sector:

§ The Classical (Traditional) Learning system: Classical learning system refers to long-established customs found in schools that society has traditionally deemed appropriate. In a traditional learning system, the teacher will talk about one subject for a set amount of time and the students listen to him.

§ The E-learning system: E-learning systems are integrated systems that support the delivery of a learning, training or education program by electronic means. It involves the use of a computer or electronic device in some way to provide training, educational or learning material.

Constraints of the classical learning system include:

1. Time: Learners are bound to a specific day/time to physically attend classes.

2. Travel Considerations: A considerable amount of time allocated to attending campus, travel time and risk of life, and travel expenses. Also, students must physically attend the courses to get credit for attendance and when combined with the cost of education, may present an issue to financially challenged students.

3. Overcrowding in Classrooms: Learning no longer become effective when there is overcrowding in classrooms resulting in unnecessary noise and distraction, inconveniences, and lack of concentration.

1.2. Objectives

The primary objective of this research is to us Php and Mysql to develop a fast, reliable, effective and efficient web-based e-learning system that will address the problems associated with the traditional learning system. Other objectives are to:

a) Develop a web-based e-learning system;

b) Implement the system in (a) above

2. LITERATURE REVIEW

2.1. The Evolution of E-Learning

As stated in [5] , people in the field of e-learning began to realize that you simply cannot put information on the web without a learning strategy for the users. In order for technology to impact positively on learning, it must fit into students live and not the other way round. As a result, e-learning was born. A few years ago, teaching and training was not done in front of a computer, but in the classroom. As technology improved, establishments began to integrate training with the computer and the field of e-learning began to take shape. In the early 1990s, many establishments were using videotape-based training for their employees [6] .

The work of [7] proposed the architecture for an integrated virtual classroom for delivering lectures, issuing and controlling assignments, with chatting, choices and forums which takes cognisance of e-learning features such as the White board, Audio and Video features. The Video feature enables one to transmit and receive video broadcasts with others in a classroom session.

The idea of putting training on video was a good idea, though it was lacking in a few areas such as: customization based on needs of users, expensive to maintain and uneasy upgrading. These videos often had limited interactions which lead to the nearly impossible task of tracking progress and assessment [8] .

Since it was obvious that video was not the best solution, a new form of training evolved, CBT or Computer Based Training. "Windows 3.1, Macintosh, CD-ROMs, PowerPoint marked the technological advancement of the Multimedia Era" [9] CD-ROMs could be cheaply produced so that the problem of checking in and out videos was eliminated. Employees were also able to simply pop in a CD to their personal computer at their desk and complete the training. Although the CD-ROM Computer-Based Training made good advancement, it still lacked the ability to track employees' performance in a central database and was also not as easy to upgrade. All these problems would disappear with the use of the Internet as a means of delivering content. The problem was, when the content was placed on the web, it was simply text to begin with and maybe a few graphics. No one really cared about the effectiveness of this new medium—it was just really according to Cooke, 2004.

One of the first innovations in actual e-learning was the LMS or Learning Management System. The first Learning Management Systems (LMS) offered off-the-shelf platforms for front-end registration and course cataloging, and they tracked skills management and reporting on the back-end [5] . This enabled schools and companies to place courses online and be able to track students' progress, communicate with students effectively and provide a place for real-time discussions. E-learning has come a very long way since its early days of being text- based via the Web or CD-ROM. So what does the future hold? There really is no saying where the field is headed. As long as training is continually geared towards the learners and strategies are used in the training, there is no end in sight for e-learning [10].

2.2. Technological Limitations of E-Learning

Students need necessary hardware for e-learning such as desktop or notebook computers and printers. Therefore, one of the major technological limitations of e-learning is the necessity of computer hardware and relevant resources. According to [11] , the lack of hardware to support e-learning in organizations is one of the factors why Small and Medium Enterprises are not willing to engage in e-learning to educate its employees. Hardware and other ICT resources are necessary for e-learning implementation in institutions. The Vietnam government spent large sums of money in buying ICT hardware for a college that implemented e-learning [12] . This work of [13] emphasized that in order to participate in online learning, both learners and staff need to have access to networked computers.

2.3. Personal Issues

As stated in [14] that preparation is indeed needed for newcomers as they may think that nontraditional learning such as e-learning is the same as a traditional learning environment. Besides, [15] specified that newcomers to nontraditional learning may get lost because they do not know what to do as there is no detailed guidance from the teacher. According to [14] , these newcomers need some orientation courses in order for them to get used to a nontraditional learning environment like e-learning. Therefore, it is not surprising to see newcomers needing to be psychologically prepared for the e-learning environment. It was mentioned in [16] that the lack of ICT skills is one of the barriers in e-learning training. As e-learning is the product of the advanced technology, e-learners will have to learn new skills and responsibilities related to the technology. According to [17] , technical skills could cause frustration to e-learning students due to the unconventional e-learning environment and isolation from others. Consequently, having to learn new technologies may be a barrier or disadvantage in e-learning for ICT novices. E-learning is not an easy task for many as it requires a lot of self-discipline.

2.4. Limitations of E-Learning System Compared to Traditional Learning System

Lacking physical interaction is another limitation in e-learning. The work of [18] expressed that the lack of physical interactions made e-learning students feel isolated and apprehensive. Lacking physical interaction may also affect the completion rate. From the work of [19] , it was revealed out that half of the students for an Advance Diploma in Education from the University of Ulster commented that it was rather hard to seek advice, as compared to face-to-face instruction. Physical classrooms however will enable learners to learn faster, as they can always refer to the instructors or peers for guidance. Body language is absent in e-learning. An example is when a student stated that he missed "facial and hand gestures", from which important cues can be derived [20] . The lack of physical interactions shown above will hinder the learning process as pointed out by [21] , that the omission of observation of student emotions may prevent professors or instructors from responding to student's needs.

Apart from this lack of physical interaction, e-learning is also criticized for not having facilities like traditional campuses: internship, volunteer opportunities, access to physical library, book stores, career and development counseling. Some learning institutions tried to provide these facilities but they were too limited [20] . He further pointed out problems such as budget, compatibility of software, and college policies that hindered the development of integrated supporting systems.

E-learning may not be suitable for certain groups of learners, especially science students who need extensive physical science laboratory experiment. UCLA's School of Dentistry spent around US$750,000 to develop their online courseware but later found out that the prospective customers would rather spend more for traditional classroom-based lectures. This may be due to the fact that these students need to carry out a lot of laboratory experiments in order to deepen their skills and knowledge, and this may be hard to achieve through e-learning laboratory simulations.

Difficulty in teaching in an e-learning environment is another issue, as instructors may not be able to teach well. Moving into e-learning is difficult for instructors who are already familiar with the traditional teaching environment. This is because the e-learning teaching environment is new and the e-learning technologies are developing and changing rapidly.

2.5. Design Limitations

Poor design of the e-learning courseware is a major issue for learners and e-learning providers, as pointed out by [21] . A poor design "gave users a feeling of being stressed and badly treated by the system". They further said this causes users to feel frustrated and eventually stop learning. Courseware design should be tailored to the needs of the learners: it should be easy to use and students should have easy access to guidance and information. According to [22] it was stated that designing the e-learning courseware is very demanding, as it should not be limited to just content and should include other components to enhance learning. In a nutshell, the poor usability of the online course will inhibit the learner's ability to acquire knowledge.

2.6. Limitations A Web-Based E-Learning System Experiment

In the work of [23] , they proposed a simple and easy to use e-learning system that is used currently in Qatar University by both students and teachers of the department of Computer Science and Engineering. The system is fully web-based and can run on any platform and compatible with most known browsers. In addition to its ability to manage electronic tutorials, the system uses an intelligent algorithm to generate quizzes and exams with different levels of difficulties according to the desire of the potential users. This algorithm adjusts the pre-set difficulty levels of the questions based on the feedback of the students. The students can see the solutions of their exams along with their marks immediately after submitting their answerers. The system reduces the constraints of the classical education system and save time and resources. In fact, teachers can set their courses, exams and quizzes at their convenient time and can track the activities of their students and guide them to acquire correctly the requested knowledge without be obliged to be physically in the university campus.

Similarly, the students can use the system freely and independently from their labs and home through the web. A discussion tool is also provided to allow teachers and students to interact together at a specific time. A bulletin board is also used to make some announcements to students about their courses and quizzes and exams. A questionnaire was prepared and distributed to the potential students registered in the system to obtain some feedback about the Web Based Training (WBT) and quizzes proposed by their system. Sixty students have been randomly selected to use as a testing pool. A group of 30 students have used the hard paper exam and another similar group has used the on-line system. Students had to obtain a grade of 60% to pass the quiz. After taking the quiz, all the students had to fill in a questionnaire with general and specific questions related to the method of testing. Finally, the majority of the students trust more a computer-based evaluation system than classical methods. Some students did comment on the absence of printed copies of their answers and the fact that they could not compare their answers with the correct results after the quiz [23] .

3. SYSTEM METHODOLOGY

The proposed system is a web application which will give learners the advantage of learning independently from any location at any time. Through a web interface, students will be able to not only learn the lessons anywhere at any time but also practice at leisure pace, thus overcoming the limitation imposed by time and space. They can also pause learning sessions at their convenience. Students can learn even at home without travelling to take lectures. With the use of a web-based e-learning system, student can learn from any location, simply by using their computers connected to the net along with the appropriate systems and tools. The system will be designed such that teachers can add, update and delete course materials using the web without being physically present at the learning center. The web-based e-learning system will allow students to learn independently at any time and from any location. Therefore, there is no limit to the number of students that can be taught at a time. Students can also view the result of a quiz, practice question, etc, after each test independent of the number of students involved.

3.1. Proposed Design of the E-Learning Platform

The system interface is entirely web-based, and is fully compatible with almost all well-known browsers. The system environment consist of the following: The first one is the client side application, which was developed with HTML and JavaScript. The second is the server program, which was developed with PHP. PHP acts as a gateway between the database and the clients. Lastly, the database management system (DBMS) which keeps and classifies all the questions is Mysql. Mysql is a relational database system. It uses Structured Query Language which is the most common database language. MySQL content can be managed via a command line or a web interface. It is open source and therefore free.

3.2. Data Flow Diagram (DFD) of the Proposed System

Data flow diagrams (Figures 1-3) define the functionality of a system. It describes the source from which data is coming, the change that occurs in the system and the source at which it ends. It is a representation of data flow through a system during which it is processed as well.

3.3. Database Table Design

Below are the table names that were created during the development of the system and their attributes.

3.3.1. Add Lessons

Table 1 stores the details of any lesson that will be added by a lecturer using the system

3.3.2. Administrator

Table 2 stores the details of an administrator.

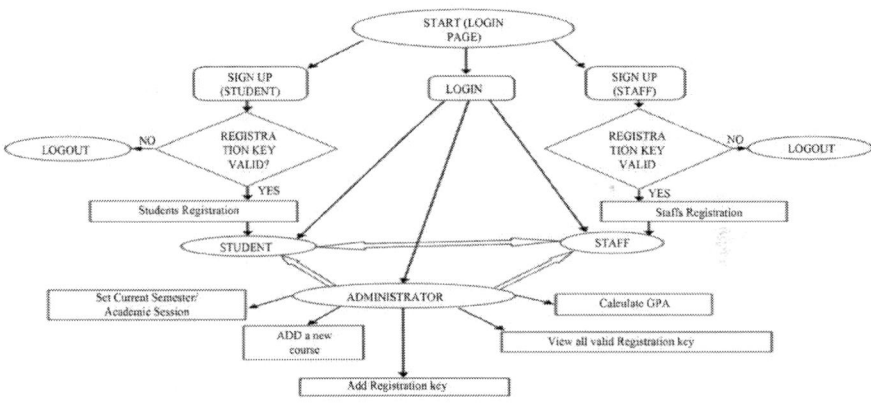

Figure 1. DFD 1 of the web-based e-learning system.

3.3.3. Administrator Check

Table 3 stores the changes made by an administrator. The administrator sets the current semester and academic session.

3.3.4. All Courses

Table 4 consists of all the courses offered in the department; the course code, course title, units and the level in which each course is taken.

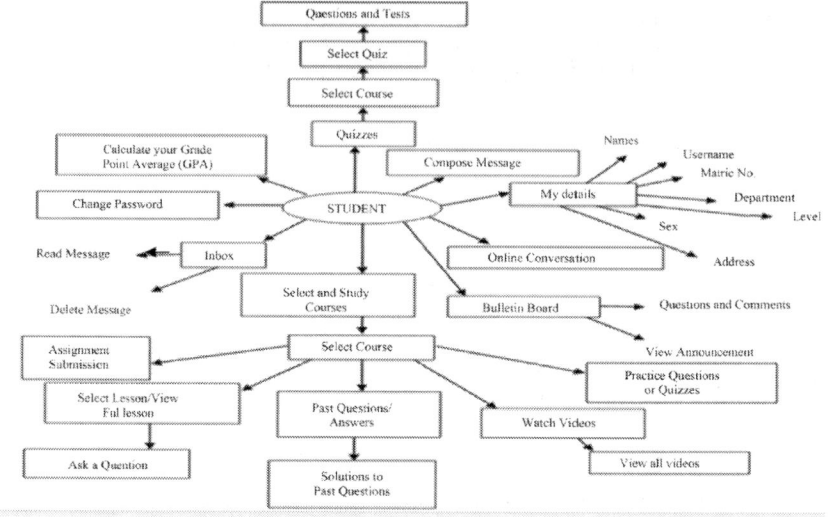

Figure 2. DFD 2 of the web-based e-learning system.

3.3.5. Lecturer Course

Table 5 stores the username of any teacher/staff managing a particular course.

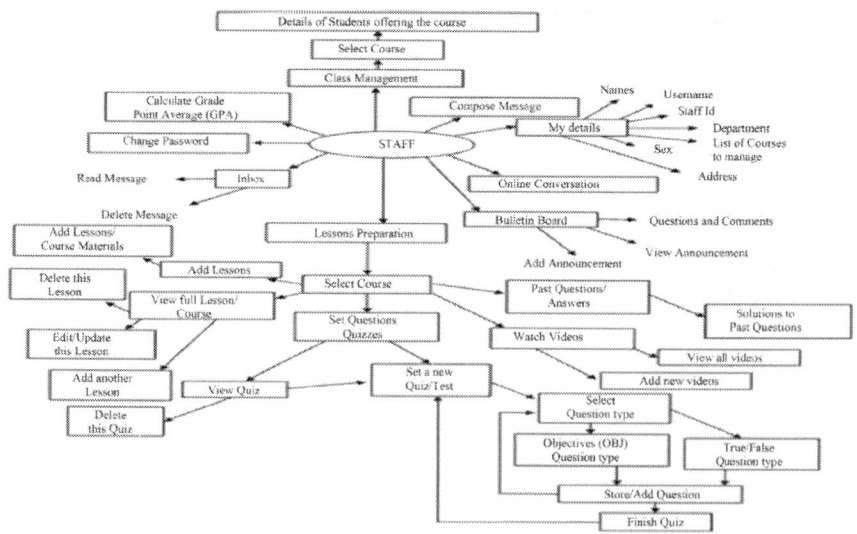

Figure 3. DFD 3 of the web-based e-learning system.

Table 1. Add lessons table.

Column Name	Primary key	Data type	NOT NULL	AUTO INC
ID	Yes	INTEGER	Yes	Yes
Course code		VARCHAR (45)	Yes	
Course title		VARCHAR (4500)	Yes	
Units		INTEGER	Yes	
Lecturer		VARCHAR (450)	Yes	
Topic		TEXT	Yes	
Lesson		TEXT	Yes	
File		BLOB	Yes	

Table 2. Administrator table.

Column Name	Primary key	Data type	NOT NULL	AUTO INC
ID	Yes	INTEGER	Yes	Yes
Username		VARCHAR (45)	Yes	
Password		VARCHAR (45)	Yes	
First name		VARCHAR (45)	Yes	

Table 3. Administrator check table.

Column Name	Primary key	Data type	NOT NULL	AUTO INC
ID	Yes	INTEGER	Yes	Yes
Semester		VARCHAR(45)	Yes	
Academic session		VARCHAR(45)	Yes	

Table 4. All courses table.

Column Name	Primary key	Data type	NOT NULL	AUTO INC
ID	Yes	INTEGER	Yes	Yes
Course code		VARCHAR (45)	Yes	
Course title		VARCHAR (4500)	Yes	
Units		INTEGER	Yes	
Level		INTEGER	Yes	
Semester		VARCHAR (45)	Yes	

Table 5. Lecturers course table.

Column Name	Primary key	Data type	NOT NULL	AUTO INC
ID	Yes	INTEGER	Yes	Yes
Username		VARCHAR (45)	Yes	
Course code		VARCHAR (45)	Yes	

3.3.6. Login

The login Table 6 stores the details of a student during his/her registration.

Table 6. Login table.

Column Name	Primary key	Data type	NOT NULL	AUTO INC
ID	Yes	INTEGER	Yes	Yes
Username		VARCHAR (45)	Yes	
Password		VARCHAR (45)	Yes	
First name		VARCHAR (45)	Yes	
Middle name		VARCHAR (45)	Yes	
Last name		VARCHAR (45)	Yes	
Department		VARCHAR (45)	Yes	
Level		VARCHAR (45)	Yes	
Address		VARCHAR (4500)	Yes	
Sex		VARCHAR (45)	Yes	
Matric No.		VARCHAR (45)	Yes	

3.3.7. Message

Table 7 stores all the messages sent by those using the system along with the recipient username and names of the sender.

Table 7. Massage table.

Column Name	Primary key	Data type	NOT NULL	AUTO INC
ID	Yes	INTEGER	Yes	Yes
Username		VARCHAR (45)	Yes	
Status		VARCHAR (45)	Yes	
First name		VARCHAR (45)	Yes	
Last name		VARCHAR (45)	Yes	
subject		VARCHAR (60)	Yes	
Main body		VARCHAR (5000)	Yes	
File		VARCHAR (45)	Yes	

3.3.8. Registration Key

The registration key Table 8 stores all valid registration keys and changes the value of the status if a particular key had been used.

Table 8. Registration key table.

Column Name	Primary key	Data type	NOT NULL	AUTO INC
ID registration key	Yes	INTEGER	Yes	Yes
Reg1		VARCHAR (45)	Yes	
Status		VARCHAR (45)	Yes	

3.3.9. Set Question

Table 9 stores the questions and quizzes set by staffs/teachers.

Table 9. Set question table.

Column Name	Primary key	Data type	NOT NULL	AUTO INC
Id	Yes	INTEGER	Yes	Yes
Course code		VARCHAR (45)	Yes	
Quiz what		VARCHAR (45)	Yes	
Question		VARCHAR (5000)	Yes	
A		VARCHAR (5000)	Yes	
B		VARCHAR (5000)	Yes	
C		VARCHAR (5000)	Yes	
D		VARCHAR (5000)	Yes	
Correct option		VARCHAR (45)	Yes	
Question No.		VARCHAR (45)	Yes	

3.3.10. Staff Login

The staff login Table 10 stores the details of a staff during his/her registration.

Table 10. Staff login table.

Column Name	Primary key	Data type	NOT NULL	AUTO INC
ID	Yes	INTEGER	Yes	Yes
Username		VARCHAR (45)	Yes	
Password		VARCHAR (45)	Yes	
First name		VARCHAR 45)	Yes	
Middle name		VARCHAR (45)	Yes	
Last name		VARCHAR (45)	Yes	
Title		VARCHAR (45)	Yes	
Department		VARCHAR (45)	Yes	
Address		VARCHAR (4500)	Yes	
Sex		VARCHAR (45)	Yes	

4. SYSTEM DESIGN AND IMPLEMENTATION

Macromedia Dreamweaver was used in the design of the web pages, PHP (Hypertext Preprocessor) as the scripting language, MySQL serve as the database and Apache as the web server. Since it is a web-based application, to implement the system, the website has to be hosted to make it accessible online.

4.1. Development Tools

The development tools used are:

- HTML (Hypertext Markup Language)

- PHP (Hypertext Preprocessor)

- Javascript

- CSS (Cascading Style Sheets)

- MySQL (RDBMS)

4.2. System Requirements

The system requirements are classified into two:

- Hardware requirements

- Software requirements

4.2.1. Hardware Requirements

The following are the minimum computer hardware requirements for effective running of the web-application:

- A full computer system (keyboard, mouse and monitor)

- Minimum of 512 MB RAM

- Minimum of 50 GB hard disk size

- Minimum of 1.60 GHz Processor speed

- An Uninterrupted Power Supply (UPS)

- A stabilizer

4.2.2. Software Requirements

For this web-application to function efficiently, the following software needs to be running on the computer system:

- A Wamp server/Apache server

- A web browser (Mozilla Firefox, Google Chrome, Internet Explorer etc)

- Relational Database Management System (RDBMS): MySQL.

4.3. Using the Web-Based E-Learning System

4.3.1. Login Page

On the login page shown in Figure 4, you can login to the system as a student, staff or an administrator depending on your username and password. On that same page (the login page), you can sign up as a student or staff if you are a new user.

4.3.2. Students Home Page

This page shows the home page for a student who had logged into the system (see Figure 5).

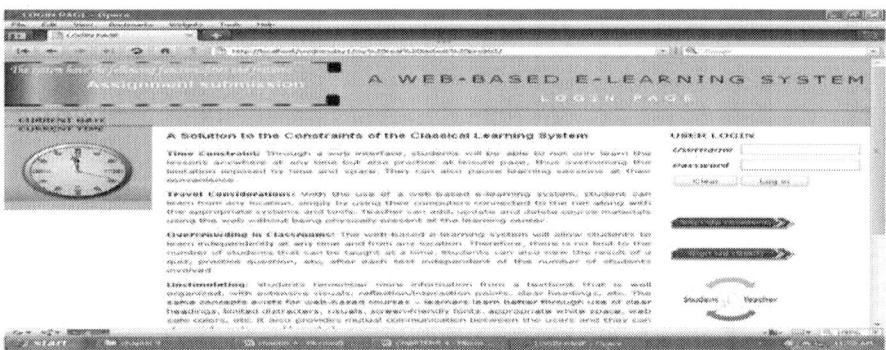

Figure 4. Login page.

4.3.3. Staff Home Page

This shows (Figure 6) the home page for a staff who had just logged into the system.

4.3.4. My Details Page

Here, both staffs and students will be able to view their personal details. The details include: Names, Matric no, Department, Sex, Level, Username, Address, etc. Staffs are also able to see the list of course they are managing.

4.3.5. Inbox Page

On this page (Figure 7), staffs and students can check messages sent to them by their colleagues and friends. Staffs can send message to students and vice-versa. Unread messages are bold compared to those ones that have been read.

4.3.6. Inbox (Message)

This page (Figure 8) allows users (staffs and students) to read the body of the message sent to them.

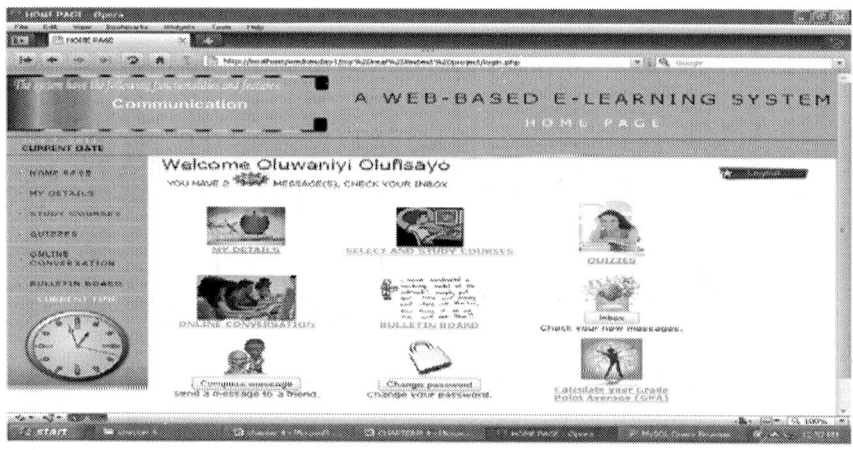

Figure 5. Student home page.

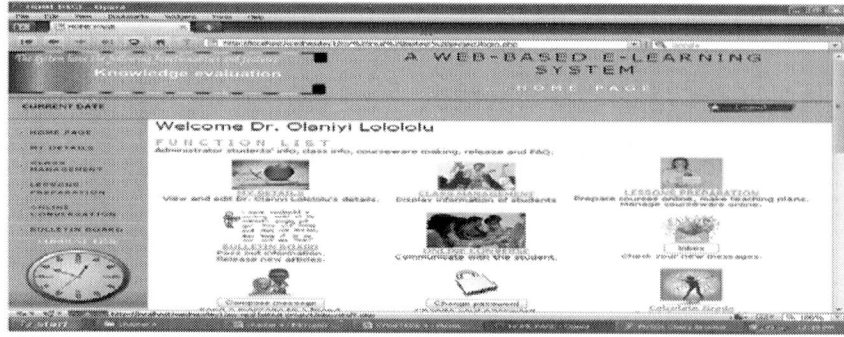

Figure 6. Staff home page.

4.3.7. The Bulleting Board

The bulletin board (see Figure 9) allows you to check information and general announcements within the system. Staffs are capable of adding announcements.

4.3.8. Online Conversation

Here, staffs and students can express their feelings and post comments which will be seen by anyone in the system (see Figure 10).

Figure 7. Inbox page.

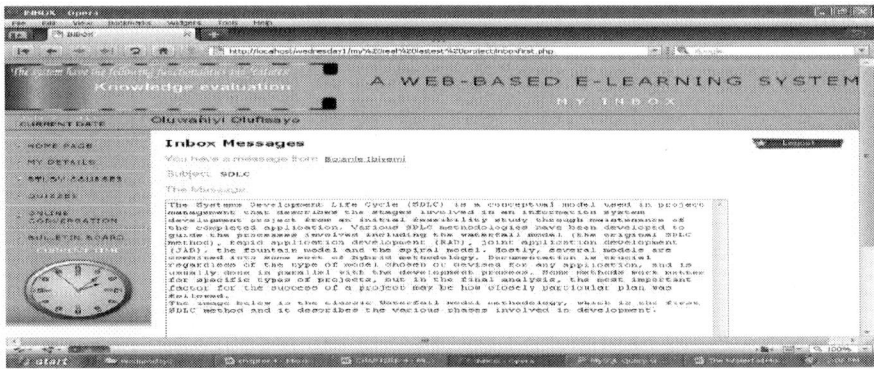

Figure 8. Inbox (message).

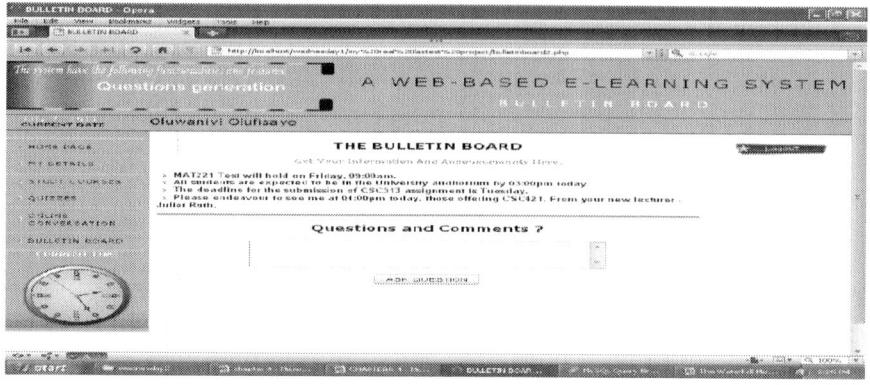

Figure 9. The bulletin board.

4.3.9. Select Course

This page (see Figure 11) allows a student to select a particular course he/she wishes to study from a drop down list.

4.3.10. Study Course Page

This page (see Figure 12) allows a student to view full lesson/course simply by clicking on any button on the page. Students can submit assignments, take practice questions and quizzes, view past questions and answers and watch videos.

4.3.11. Grade Point Average Page

Students, staffs and administrator can calculate grade point average on this page (see Figure 13).

4.3.12. Class Management Page

On this page (see Figure 14) the details of students offering a particular course is displayed, their names, matric number, department, level, username and the overall percentage score in tests/quizzes for a particular course.

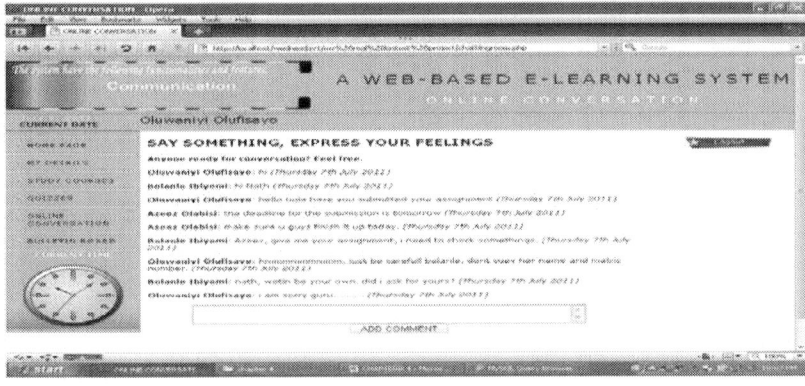

Figure 10. Online conversation.

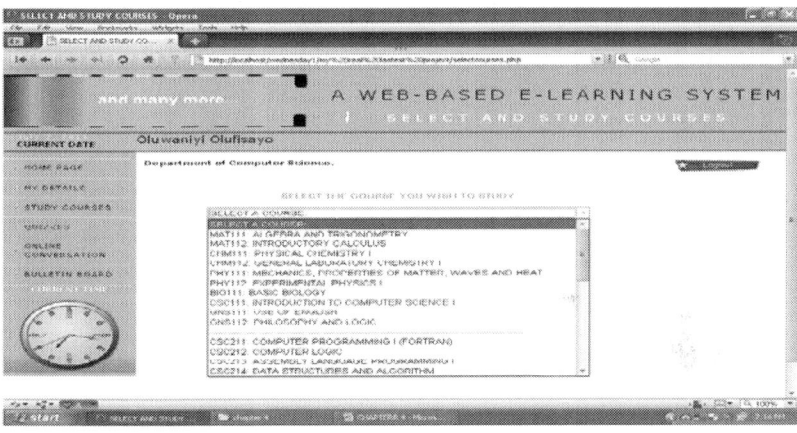

Figure 11. Select course.

Figure 12. Study course page.

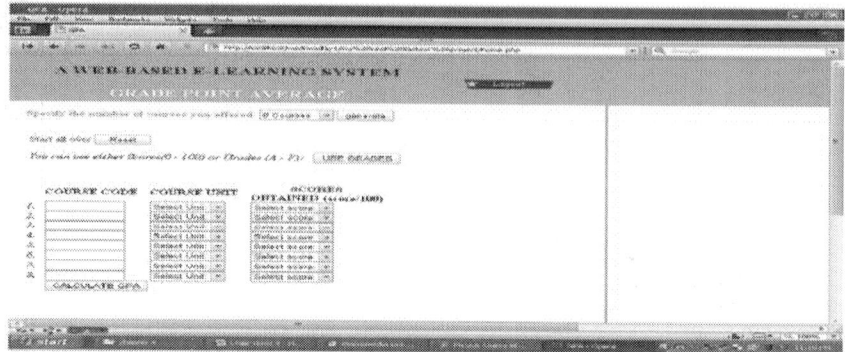

Figure 13. Grade point average page.

4.3.13. Administrators Page

This page (see Figure 15) shows the main page of an administrator. On the page the administrator can set the current semester and academic session, add a new course, add registration key, view all valid registration keys and calculate grade point average (GPA).

5. CONCLUSION AND RECOMMENDATION

5.1. Conclusion

The web-based asynchronous e-learning system will reduce the constraints of the classical learning system and save time and resources. Course lecturers can set their courses, tests and quizzes at their convenient time and can track the activities and performance of their students and guide them to acquire knowledge without being obliged to be physically present on the institution campus. This system is no doubt a solution to the constraints of the classical learning system.

5.2. Recommendation

Educational institutions need to be awaked and embrace e-learning system in order to meet the growing need for education and technology. It would be worthwhile if education agencies collaborate with institutions to fund researches on e-learning and virtual classroom instead of spending money in constructing classrooms that will never be adequate for the ever-increasing population. Government should to come up with legislations and guidelines for the implementation of e-learning and virtual classrooms so that the certificates obtained through e-learning can be universally acceptable.

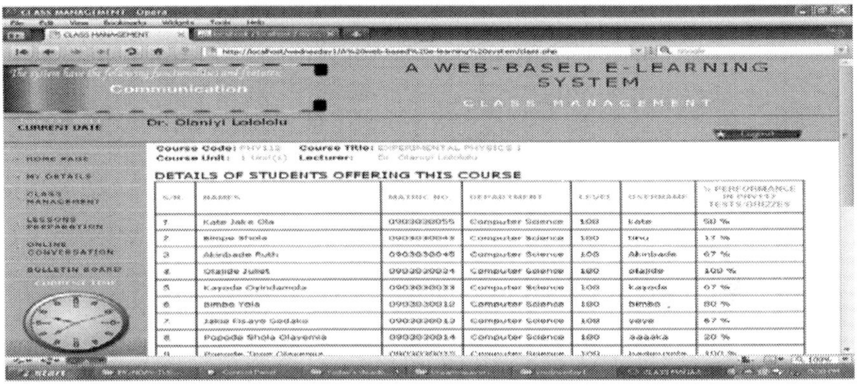

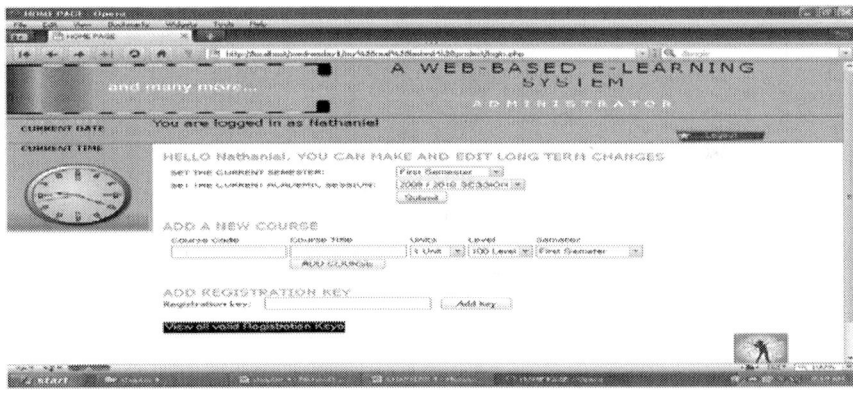

Figure 14. Class management page.

Figure 15. Administrator page.

CITE THIS PAPER

Nathaniel OlufisayoOluwaniyi,Babajide OlakunleAfeni,Olusegun Olayinka Lawal, (2015) Development of an Asynchronous Web Based E-Learning System. *Journal of Computer and Communications*

REFERENCES

1. Ramim, M. and Levy, Y. (2006) Securing e-Learning Systems: A Case of Insider Cyber Attacks and Novice ITmanagement in a Small University. Journal of Cases on Information Technology, 8, 24-34.

2. Alias, N.A. and Zainuddin, A.M. (2005) Innovation for Better Teaching and Learning: Adopting the Learning Management System. Malaysian Online Journal of Instructional Technology, 2, 27-40.

3. Karim, M.R.A. and Hashim, Y. (2004) The Experience of e-Learning Implementation at the Universiti Pendidikan Sultan Idris, Malaysia. Malasian Online Journal of Instructional Technology, 1, 50-59.

4. Nichols, M. (2003) A Theory for E-Learning. Journal of Educational Technology and Society, 6, 1-10.

5. Nor, A. and Ahmed, M. (2005) Innovation for Better Teaching and Learning: Adopting the Learning Management System. Malaysian Online Journal of Instructional Technology, 2, 27-40.

6. Ajayi, I.A. (2008) Towards Effective Use of Information and Communication Technology for Teaching in Nigerian Colleges of Education. Asian Journal of Information Technology, 7, 210-214.

7. Zhang, D., Zhou, L., BrIggs, R. and Nunamaker, J. (2006) Instructional Video in e-Learning: Assessing the Impact of Interactive Video on Learning Effectiveness. Information and Management, 43, 15-27.

8. Douglas, I. (2001) Instructional Design Based on Reusable Learning Object: Applying Lessons of Object-Oriented Software Engineering to Learning System Design. Proceeding of ASEE/IEEE Frontiers in Education Conference, Reno, 2001, Vol. 3: F4E1-F4E5.

9. Wood, R. and Ashfield, J. (2008) The Use of the Interactive Whiteboard for Creative Teaching and Learning in Literacy and Mathematics: A Case Study. British Journal of Educational Technology, 39, 84-96.

10. Evans, C. and Fan, J.P., (2002) Lifelong Learning through the Virtual University. Campus-Wide Information Systems, 19, 127-134.

11. Leonard, B.., Akio, K., Arjan, D. and Giuseppe, D.M. (2006) A Web-Based e-Learning System for Increasing Study Efficiency by Stimulating Learner's Motivation. Published Online: 14 November 2006.

12. Kibble, J.D. (2007) Effective Use of Course Management Systems to Enhance Student Learning: Experimental Biology 2007. Adv. Physiology Education, 31, 377-379.

13. Hosam, F.E., Ahmad, M.H., Jehad, M.A., Fayed, F.M.G. and Samir, A.E. (2006) A Web-Based E-Learning System Experiment. International Journal of Computing and Information Sciences, 4.

14. Teh, G.P.L. (1999) Assessing Student Perceptions of Internet-Based Online Learning Environment. International Journal of Instructional Media, 26, 397-402.

15. Lee-Post, A. (2009) E-Learning Success Model: An Information Systems Perspective. Electronic Journal of e-Learning, 7, 61-70.

16. Leonard, B., Akio, K., Arjan, D. and Giuseppe, D.M. (2006) A Web-Based e-Learning System for Increasing Study Efficiency by Stimulating Learner's Motivation. Published Online: 14 November 2006.

17. Al-Adwan, A. and Smedly, J. (2012) Implementing E-Learning in the Jordanian Higher Education System: Factors Affecting Impact. International Journal of Education and Development using Information and Communication Technology (IJEDICT), 8, 121-135.

18. Trentin, G. (2006) The Xanadu Project: Training Faculty in the Use of Information and Communication Technology for University Teaching. Journal of Computer Assisted Learning, 22, 182-196.

19. Toomey, R., Priestly, I., Norman, A. and Mahony, G.B. (1998) Effective Teaching and Learning in a Simulated Environment: A Higher Education Case Study. Journal of Hospitality & Tourism Education, 10, 28-32.

20. Obasa, A.I., Eludire, A.A. and Isaac, M. (2011) The Architectural Design of an Integrated Virtual Classroom System. Research Journal of Information Technology, 3, 43-48.

21. Mahanta, D. and Ahmed, M. (2012) E-Learning Objectives, Methodologies, Tools and Its Limitation. International Journal of Innovative Technology and Exploring Engineering (IJITEE), 2, 46-51.

22. Arkorful, V. and Abaidoo, N. (2014) The Role of e-Learning, the Advantages and Disadvantages of Its Adoption in Higher Education. International Journal of Education and Research, 2, 397-410.

23. Yannis, P. (2004) An Approach for Managing the Evolution of Web-Based Educational Applications. International Journal of Web Engineering and Technology, 1, 148.

CHAPTER 12

Personalization and User Modeling in Adaptive E-Learning Systems for Schools

Todorka Glushkova[1]

[1] *Plovdiv University "Paisii Hilendarski", Bulgaria*

Abstract

The manuscript presents a model for the personalization of e-Learning systems in secondary schools. Approaches are discussed about the implementation of this model by the application of the SCORM-standard, ITL (ITL-Interval and temporal logic), policies, etc. Comments on the possibilities for increasing the relevance of e-Learning systems in the real classroom environment schools are also included.

Keywords

E-Learning, User Modelling, Personalization, Adaptation, Interaction, SCORM, ITL

1. INTRODUCTION

Dynamically changing realities in modern society require more dynamic and adequate changes in education, which is inherently conservative. Every innovation and change in the traditional school system requires a long time for synchronizing the legal basis, approbation, and implementation in school practice. This generates a continuous delay in the fast increasing requirements to the educational system, which makes it difficult to meet public expectations. The world community sees a way to solve this problem through the application of ICT and e-Learning in the educational process. Environments that offer a variety of teaching materials and services to different user groups, such as students, parents, learners, employees, etc., are created. As a rule, all these systems are developed faster and cost much cheaper than their traditional equivalents. They enable the implementation of new and different approaches from the traditional and provide solutions as well as access to educational materials within the process of learning. The standardization of individual modules and processes bring order to the variety of systems for computer training and make it possible to use them independently because of the software and hardware platform features. All this seemed to solve the problems of didactic theory and practice, but the reality is different. There is a delay in the process of implementation and actual use of these systems despite the rapid development of information and communication technologies. Research groups from different universities and educational communities define the main reason as learning by means of ICT is an innovative process that requires in-depth research by pedagogical science and practice [1]. Psychological attitudes, motivation, and cognitive characteristics of students of different age groups are different in terms of learning processes. The didactic theory and practice for many years examine these processes and provide solutions for the improvement of the traditional forms of training, however, the mechanisms in e-Learning environments of educational portals are still not well investigated and explained. There is a gap between the expectations of developers and real results in the educational practice. The rapid development of computer science and technology for the creation of learning environments requires a high level of qualification and experience of the developers of such systems. These developers are highly qualified and highly specialized IT professionals who create systems in accordance to their abstract vision of the learning process. However, they rarely or never are pedagogical specialists and therefore, do not know the actual in-depth psychological processes of learning. This leads to the fact that institutions have learning environments that have a good software perspective quality but poor quality as a pedagogical tool. To these reasons, we can add the difference in terminology, the desire to maximize profits of software vendors, unclear criteria for evaluation of the learning process, etc. As a result, we see many factors that negatively affect the whole process and can significantly hinder it.

Despite these problems, there are many areas in which the results are very good. Summarizing the results of higher education in the analysis of the Sloan Consortium in [2], successfully use of e-Learning environments in the USA and Canada grew by 20% over the last few years. The indicators are particularly good for students who are trained in a distance form of teaching, as well as with electronic training courses that are similar in nature to the traditional teaching process. Secondary schools also have sectors that experience very good results. Examples are as follows: the using of educational environments in blended learning (classroom training and independent work) and the creation of a portfolio of students who successfully combined with project training [3].

The problem analysis allows us to conclude that the creation of training systems must comply with specifics of the particular educational institution and be developed in direct communication with educational experts as were probed directly in the real learning environment. This publication will present a model for personalizing learning systems for electronic and distance learning in secondary schools by application of didactic methodology, setting of educational goals and objectives, the motivation of the student, and his or her personal goals, plans, and ambitions.

The structure of the manuscript corresponds to the described methodology. In section 2, "Interaction and Adaptation" discusses various aspects of interactivity and adaptability connected with the personalization of the learning process and provides access to educational resources. Here, the adaptive levels of the system in horizontal and vertical plans are reviewed.

Section 3, "Adaptive levels and interaction Student-Learning system", describes the three adaptive levels and some mechanisms for their implementation of the SCORM-based e-learning system.

Section 4, "Personalization and User modelling", are connected with the opportunities for development of the personalization and UM on different adaptive levels according to the described methodology. An algorithm is proposed for the implementation of the model in the e-Learning environment.

The results of the partial implementation of the proposed model of e-Learning in secondary schools are encouraging. Work on the realization of the full adaptive model continues.

2. INTERACTION AND ADAPTATION

According to the definitions in [4], [5], and [6] e-Learning is a computer and an internet-based learning, in which the delivery of electronic learning resources is carried out on the principles of dynamic interaction with the educational system and the other participants in the learning process, according to didactic set goals and objectives and according to the characteristics of the course and the personal characteristics of the student. Based on this definition, the team of University of Plovdiv, together with partners from the Institute of Information Technologies (BG), University of Limerick (Ireland), De Montfort University in Leicester (UK), Humboldt University (Germany), Secondary school in Brezovo (BG), etc., developed a system for electronic and distance training (DeLC[1] -). As part of this project, an environment is developed for e-Learning and distance training for secondary schools. In order to minimize the problems mentioned above, we chose a methodology by which, together with pedagogical specialists, creates step by step the different prototypes of the system and tested them directly in the real learning environment (Figure 1).

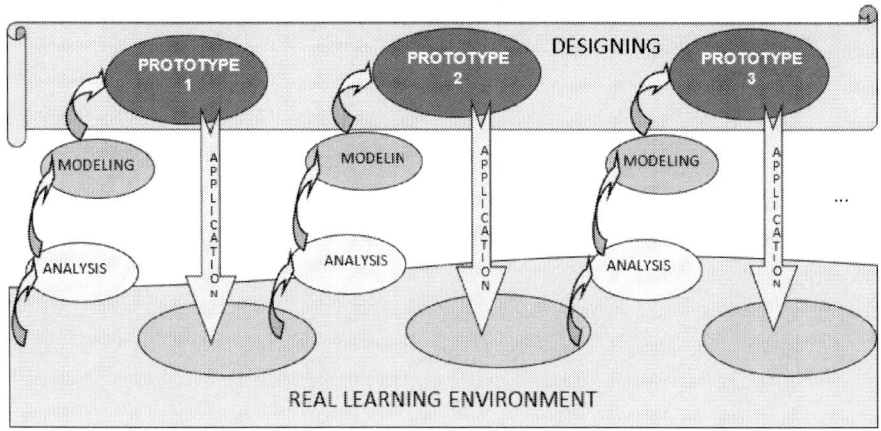

Figure 1. Iterative model of research methodology

To create an interactive and adaptive system that meets these requirements, it is necessary to model the different interactive levels and adaptive aspects. We accept the definition in [7] and define it as a dialog between users and the learning system and will view it at the following three levels:

- Standard Experience – the physical structure and hierarchy of the learning content remains unchanged;

- Personal Experience – the hierarchy of content changes and adapts to the user's behaviour and selections;

- Open Experience – an open and live system with continuous engagement between the producer, user, and message.

Within these levels, there are six categories of interactivity: Feedback; Control; Productivity; Creativity; Communication and Adaptation. Ensuring interactivity for each level and related categories requires the development of a comprehensive adaptive model to provide personalization of the training through: the basic knowledge of the student; his plans and purposes; his cognitive characteristics; his preferences and habits; emotional profile, and so on (see [8]).

According to various aspects of the application and the use of e-Learning systems, the adaptability can be defined differently. We will define adaptability as a feature of the training system to be adapting itself and changed according to the requirements and the characteristics of the users before and during use of it. The main elements of the adaptive model are "condition-action" rules that change the parameters of the environment and realize the adaptation to a user's knowledge, goals, abilities, preferences, etc. The different methods and frameworks for creation of adaptive models are as follows:

- Rule-based – we look at these systems from some aspects: as a declarative interpretation of rules; as a hybrid representation based on logical deduction; as a users' stereotypes; as an overlay model of connecting and co-interacting with the model of the relevant applied area. Presenting the cognitions is connected with the accommodation of the system conventions, the attitude and the convictions of the users, and the stereotypes and the user groups that can be activated dynamically with the particular conditions, etc.

- Frame and Network-based – these models are associated with figuring the sciences as interrelations between separate facts of semantic net and frame structures. It can be used successfully on a small applied domain that can be easily identified and structured.

- Supposition-based – Such systems work with a multitude assumptions of the consumer that forms on the student's knowledge base and domain independent rules. The suppositions are facts on the user that the system takes with a certain level in security and cogency. The degree of cogency in the system is being raised if the user gives a good feedback and falls if he gives a bad one. The suppositions, established on the base of the direct communication with the user, are better defined that the system is adopted on the base of logical deductions. Formally, we can differentiate the system assumptions on three categories: what the student knows; what he doesn't know; and the student's aims, tasks, and plans. The first two can be realized by stereotype and overlay models as the knowledge of students are being adapted to relevant domain ontology. The third group is part of the Goal and Task Model of the e-Learning system. The most simple is the method of linearly parameterization. It is more complex but more reliable as the model uses formulas, predicates, ITL, policies, and grid models.

- Based on statistical rules and theoretical conclusions – This model permits the adapting of rules according to the state of the entrance data. The opportunities for setting-up the adaptation are based on the information from past learning sessions.

For the development of the adaptive model, we use separate elements from each of these methods we: use the stereotypes and the overlay model at the initial determination of common behaviours rules; define the concrete dependencies from the described school subject domain, by using of domain ontologies; make a system from assumptions for the learner, based on his stereotype, cognitions, and goals; store the information from the last learning sessions and processed it statistically; and deduced abstract conclusions for the user groups and the separate learners. The implementation of the model requires the consideration of the various adaptability aspects of horizontal and vertical principles. The first one is connected with the adaptation to some personal characteristics of the student. This model of realization in this aspect is discussed in [9]. There are two types of adaptation – adaptivity and adaptability. The first allows users to use different facilities for presentation and navigation in predetermined learning content. The second level includes mechanisms for adaptation to knowledge and preferences of students dynamically in the learning process. On this basis, we will distinguish the following three adaptive levels: Elementary Adaptive Level, Static Adaptive Level, and Dynamic Adaptive Level. The first two are connected with adaptivity, and the last one – with adaptability.

In the next section, we will look at these adaptive levels. We will focus our attention on the Static Adaptive Level and will comment on some ideas for the realization of the Dynamic Adaptive Level.

3. ADAPTIVE LEVELS AND STUDENT-LEARNING SYSTEM INTERACTION

Elementary adaptive level (EAL) – the adaptation in this level is connected by the use of static user information such as type of training, grade, class, name, access to learning material (by mobile or fixed device), etc. Here, we can use a stereotype approach. The authors of educational resources develop packages of lessons, tests, etc., in accordance with government educational requirements and standards for the typical student, school subject, the class and form of training. The created educational resources are common for all students in the described groups.

We can realize an adaptation on this level by the preparation of training materials, common for all students in the phase before the start of the learning process. These characteristics prove the relationship with the first interactive level – Standard Experience, because the physical structure and hierarchy of the learning content remains unchanged. However physical and cognitive interaction occurs for the users. The student receives the entire information independently if he knows it. At this level, users are an abstract group of people with common characteristics – background knowledge, preferences, cognitive performance, and more. This personalization is the lowest formal level. Although users have their accounts in the school e-Learning environment, they can work in it only on the predetermined way for all other users in the same stereotyped group.

Static adaptive level (SAL) – this level is based on the elementary level and is directly related with mechanisms to provide adequate learning materials for individual students according to their knowledge base, personal goals, and plans.

Before we present the adaptive mechanisms of this level, we will comment on the concept of persona as an aggregated user type. The persona is a description of a fictitious learner. This description is based on different methods, including the personal experience of the teachers, hypotheses, statistical methods, and heuristic analysis [10].

Adaptability of this level is realized through the collection preparation of educational materials and services, foreseeing the actions and behaviour of the persona. The realization is based on the log-information about past interactions between the student and LMS according to the set of rules defined by the teachers. It is necessary to define the background knowledge of this student. We can determine the knowledge in different ways – by initial testing, by results from completed to this moment training sessions, etc. Based on these values, the system joins this student to the persona, who is closest to these characteristics. The system compares the individual characteristics, plans, and objectives of the student with the typical didactic aims, defined in BES. As a result, this lesson that most closely matches with the basic knowledge, goals, plans, and personal characteristics of the student that is associated with this persona starts from the Lesson repository.

Moreover, special attention will be paid to the creation of courses by pedagogical specialists in the article. In accordance with pedagogical theory, this process is cyclical and begins by placing the main didactic aims, passes through specifying learning tasks, develops the profiles of aggregated user types (personas), the establishes learning scenarios with different personas, develops prototypes of the training process, shares this prototype among specialized pedagogical community of educators, experts and heuristic evaluation of this prototype testing in real learning environment, and corrects existing errors and inaccuracies. Each stage of this process requires a qualitative evaluation and heuristic analysis of the pedagogical community and the implementation of appropriate tools in a common environment – ex. Integrated Learning Design Environment (ILDE) [11] (Figure 2).

Creation of the training course, according to the developed scenario is realized based on the selected standard for e-Learning. In developing the DeLC-system, we use the standard SCORM[2] - [12]. The team developed a special SCORM-editor for the teachers and authors of the educational content – SELBO [13].

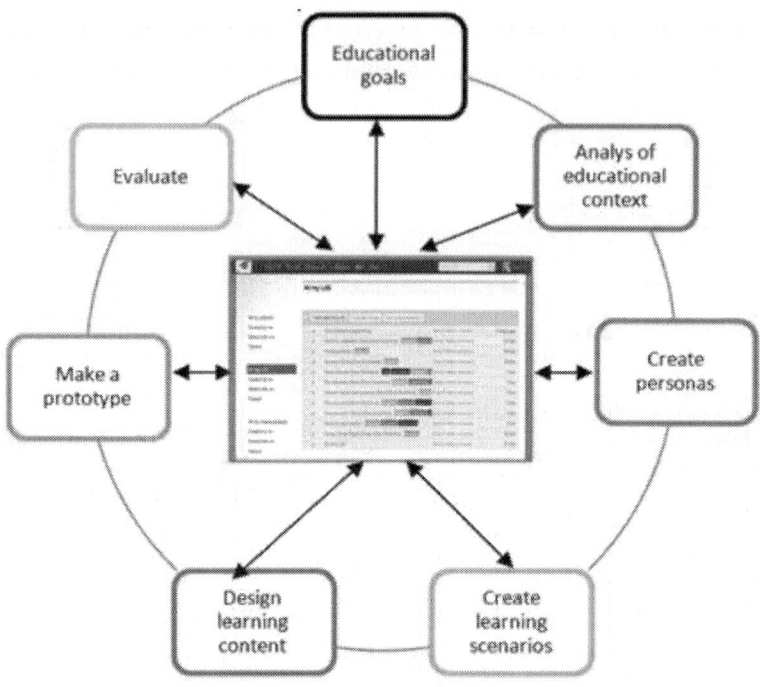

Figure 2. Creation of learning course in ILDE

We will concentrate our attention on the two basic characteristics of the lesson – the content and the structure. The content of lessons is related with specific topics that are connected with some school subject domains. The e-lesson presents a semantic structure of the knowledge that is connected with some school subjects. The formalization can be realized through the creation of ontologies in which each concept from the respective area are associated with real information resources that represent them in the lesson. According to the main characteristics of the school subject domain, the authors of e-content in accordance with didactic aims define the structure of the lesson. The didactic aims are related with type of the lesson (for new knowledge, for exercises, summary, and testing). We use Bloom's taxonomy to formalize these aims with the cognitive levels – knowledge, comprehension, application, analysis, synthesis, and evaluation [14]. The teacher could structure the lesson in different ways depending on the predefined didactic aims. We came to the conclusion that there is a correspondence between the different types of e-lessons and the cognitive levels of Bloom's taxonomy.

Therefore, we can formalize the different types of e-lessons according to didactic aims by creating standard scenarios for training and templates that describe them. The template is a combination of structure and learning scenarios.

To create electronic lessons by using algorithm requires a thorough knowledge of the standard SCORM, which creates objective problems for teachers who are not IT specialists. To partly solve this problem, we can use the SCORM Best Practices Guide for Content Developers (BPG) [15], which offers a number of basic templates and models that correspond to different educational scenarios.

In order to increase the formalization level of templates and models, we created a system for its parameterization. Thus, we received a number of different groups of templates that more fully meets the requirements and objectives of the learning process. We can use the following for the parameterization of templates:

- Number_of_SCOs – type Integer, to describe the number of SCOs[3] - in the template. If the parameter value Has_test is "yes", the number of questions Num_Quest =Number_of_SCOs – 1;

- Has_test – type Boolean to determine whether there is a final test or not in the template. Defaults to "yes" and is realized with the last SCO;

- Num_Quest – number of questions in the final test

- The ordered pair (objective_n, min_value_n) connects each target variable (objective) and the minimum value for which LMS will mark it as successfully passed (n<Number_of_SCOs);

- The ordered pair (SCOn, template_num), for each n <Number_of_SCOs and template_num <= 10, which connects each SCO with instance of the main BPG-template;

Let's define two operatios:

- Set (SCO_Number_of_SCOs (Asset k); Objective_k) – for setting values of k-th target variable for the k-th issue of the last SCO, where k <Number_of_SCOs, and

- Read (SCO_k, Objective_k) – start-up of the information SCO_k if Objective_k has a value less than the predetermined.

In dialogue with the SCORM-based authoring tool for generating of electronic lessons, the teacher will determine the values of these parameters and the system will generate the structure of the desired template. If specific values of the parameters are not mentioned, the system will get the default values. After the parameterization, the teacher will receive the parameterized template with the SCORM rules that served as a guide to the sequence of educational activities in the educational scenario depending on the behaviour of the individual student.

For example, if we get parameter values: number_of_SCOs = 10; has_test = "yes"; (objective_n; 0.75) for each n <10; (SCO_n, template_2), for each n <10; Set (SCO10 (asset_n), objective_n) for each n <10; Read (SCO_n, objective_n) for each n <10, we get the following chart describing the scenario of the lesson (Figure 3):

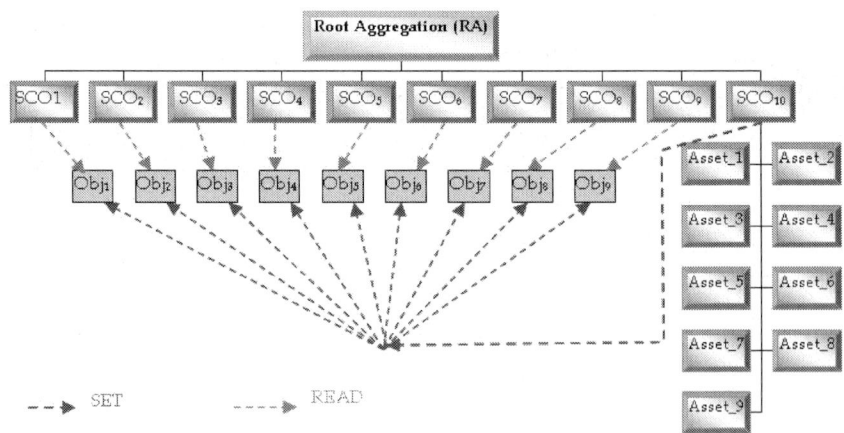

Figure 3. Parametrized Content Structure Diagram

Therefore, based on the Bloom taxonomy of didactic purposes, lesson types formalization of their structure and navigation rules can be created by the step by step algorithm for creating of e-lessons.

The teacher in some school subjects create e-lessons in a specialized development environment, which is in dynamic interaction with the respective ontology. This author's tool has to maintain information about the compulsory concepts in the relevant discipline according to Bulgarian Educational Standards. Concepts that are mandatory taught according to BES and those that are determined from the teacher for this lesson are marked with AND and those that contain additional information are marked with OR. The algorithm includes the following steps:

1. Through a dialogue with the system (e.g. personal assistant), the teacher defines the class, form of training, and didactic aims of establishing a lesson. The system filters relevant domain ontology and retrieves concepts – mandatory and complementary.

2. Depending on the subject area and selected didactic aims, the system offers the most appropriate templates that can be selected from.

3. In a step by step process, the teacher determines the values of parameters such as the number of data objects (SCOs), presence of preliminary and final test, minimum values of the target variables for their passing, number of attempts to solve the tests, etc. As a result, the system generates a parameterized template, which includes the structure of the lesson (SCORM CAM[4] -) and the rules that will manage the learning process (SCORM Sequence & Navigation Model[5] -).

4. The author connects the SCOs with nodes of the structural graph in the template of the lesson.

5. The author ccreated lesson (the system generates zip-package and imsmanifest.xml)

6. He upload created lesson in SCORM-environment of the education portal.

E-Learning resources (SCOs) are associated with concepts of relevant ontology. They are stored in some online SCOs Repository. In ontologies and related items, SCOs are presented into the development environment for creating electronic lessons [16]. The authors determine the structure of the e-lesson by using the parameterization of some basic templates. In this way, they create an instance of the template in which there are no free parameters. LMS manages educational processes and determines the training scenario in accordance with the structure of the lesson and learning scenario, which are related to the didactic aims, behaviour, and basic knowledge of the students. The e-lesson will be presented in the system as a specific instance of some basic template, which, by setting the values of parameters, is associated with specific learning resources.

One example is the Lesson "Summary on complex verb tenses» for 7th grade students for independent distance training". The didactic aim is to reach higher levels of Bloom's taxonomy – application, analysis, synthesis, and evaluation. The system offers BPG-templates №7, 8, and 10, and the teacher chooses Template 7. In this template, SCOs containing learning information are grouped in a separate Aggregation B. The student must answer questions from the preliminary test and if wrong (i.e. target variables that monitor test results are less than the minimum values), he has to become familiar with the educational content of the information SCOs. Then, he will make the final test in the last SCO. The template can be used in the creation of educational resources, which is necessary to verify and ensure a certain volume of background knowledge. It is essential to fill the gaps and to allow the student to successfully pass the final test. LMS manages the values of the variables (objectives) and only if they are larger than the specified minimum, the training is considered to be successfully completed. The teacher gives the following values of the parameters through a dialogue in a step-by-step process: Number_of_SCOs=9; Has_pre_test="yes"; Num_Quest=9; Has_post_test="yes"; $(Obj\_n,1)$ – i.e., answered correctly for all questions from 1 to 9; $(Obj\_n; 0,75)$ – gave a very good answer to the questions for $\forall n \in [10,18]$; (SCOn, pattern_2) for $\forall°n \in [3,11]$; Set(SCO1 (Asset_k); Obj_k) for $\forall k \in [1,9]$; Set (SCO2 (Asset_k); Obj_k) for $\forall k \in [10,18]$; Read (SCO (k+2), Obj_k) for $\forall k \in [1,9]$. The system generates a CAM- model and S & N rules. The teacher writes the test questions and puts the SCOs in Aggregation B. These data objects (SCOs) include both basic information on various complex verb tenses as well as tasks and exercises for students who will have to pass successively through the levels «application» – «analysis» – «synthesis» – «evaluation». The teacher makes a SCORM-package of the lesson and uploads it in the SCORM-based school education portal (http://sou-brezovo.org). These characteristics prove the relationship with the second interactive level – Personal Experience – because the hierarchy of content changes and adapts to the user's behaviours and selections.

Dynamic adaptive level (DAL) – This level is related to the dynamic interaction between students and the system during the training process (in run-time). After selecting the most appropriate e-lesson from the Lesson DB, the LMS starts the learning process according to the training scenario. The learning scenario is realized by a sequence of actions that is previously defined by the author of the lesson. The system observes the intermediate results during the training and information from the already completed training sessions. Based on this information, the LMS adapts itself dynamically to the changing characteristics of the learning environment as it generates new "condition-action" rules and either continues the training process or stops it. If the parameters are not appropriate, the system has to choose and to start a new and more appropriate e-lesson.

We are convinced that in the process of dynamic interaction between the learners and the training system, it is essential to use intelligent agents who interact with the system and with each other to provide a flexible change of training scenarios depending on the behaviours and actions of the individual student. For the managing of the dynamic adaptation of LMS, we can use Interval temporal logic (ITL) and policies.

Morris Sloman in [17] defines the policies as a set of rules for activating different states and actions, depending on the behaviour of the consumers or the current state of the system. There are different techniques to formalize the policies – graphical modelling, using the object-oriented methods for defining of policies, etc. We will use the opportunities provided from ITL [18] as it builds on a classical logic tier and allows to describe dynamic processes in the course of their implementation. It is a flexible notation for handling events that varied in time intervals, allows series, and parallel compositing using a well-defined mathematical proof system. ITL includes four components – logic tier, temporal structures, conditions, and intervals.

Classical logic manages variables, constants, functions, and predicates. If we want to describe the dynamic processes, it is necessary to add temporal structures as skip, chop, and chopstar. The states are specific transmission of values to the observed variables and the intervals are sequences of states.

We will describe the next three sets: S-set of students, O-set of available objects or resources, and A-actions that can be performed with these resources. Then, we can introduce the user authentication as one of the Boolean variables:

- Autho+(S,O,A)Autho+(S,O,A) – Positive identification of the user S, who has right to use the resource O by performing action A. For example Autho+(Ivan, Lesson1, Read) Autho+(Ivan, Lesson1, Read) or Autho+(Ivan, Test1, Write);Autho+(Ivan, Test1, Write);

- Autho−(S,O,A)Autho−(S,O,A) – Negative identification – the user S refusal to use the resource O by performing action A. For example Autho− (Ivan,Lesson1, Write).Autho− (Ivan,Lesson1, Write).

Upon the initial start-up of the system, these variables have a default value of "false". The mathematical model of **Autho** is a matrix with 3 columns – users, objects, actions, and n-number of lines for all users in the system. The access to resources will be allowed if they satisfied certain "condition-action" rules of the type: F→W i.e., F always followed by W in the final state of the observed subinterval. According to this definition, the Access Rules take the following form: F→autho$^+$(S, O, A) – rule for positive identification and F→Autho$^-$(S, O, A) – rule for negative identification. For example: If in the initial step the access was denied, but in the next moment, it is authorized in the duration of 10 time units then: $((\text{Autho}^-(S, O, A) \wedge \text{skip}) \vee (\text{Autho}^+(S, O, A) \wedge \text{len} <= 10) \rightarrow \text{Autho} +(S, O, A)$.
If two users M and N are grouped and one of them has access, then the second one also receives access: $\text{In}(M,N) \wedge \text{Autho}^+(M,O,A) \rightarrow \text{Autho+}(N,O,A)$.

The Access Rules determine whether the particular user is entitled to access this learning resource or service. To realize the access itself, the management passes the Implementation Rules, which has the following more general form: F→Autho(S,O,A). There are two alternatives in access: Open Access and Restricted Access. Open Access has low security – i.e., if access is not prohibited, it is allowed: $\neg Autho^-(S, O, A) \rightarrow Autho(S, O, A)$. Restricted Access means the system checks whether access is allowed and it has meanwhile been prohibited i.e., $(Autho^+ (S, O, A) \wedge \neg \, Autho- (S, O, A)) \rightarrow Autho(S, O, A)$.

Another way to access learning resources is the delegation of rights to the unauthorized user. For example, the teacher gives access rights to other teachers for reading a lesson: Teacher $(S, Lesson) \rightarrow$ Candeleg+ $(S,\_,Lesson,Read)$. The rules for delegating access, which author A1 gives teacher T2 to make corrections in Lesson1 is: $(Autho(A1,Lesson1, Write) \wedge Candeleg(A1,T1,Lesson1,Write)) \rightarrow Autho(T1,Lesson1, Write)$.

The policy P is a collection of rules: Unexpected text node: 'w 'Unexpected text node: 'w ', where w is the initial state, w' is the final state, and $\wedge r_i$ is a conjunction of intermediate states. For example, the policy for Author of Lesson1 (Author), teacher, who use this Lesson1 (Teacher), and student (Student) is:

$$P1 \cong \big(\big(Author(S, \; Lesson1) \rightarrow Autho^+ (S, \; Lesson1, Read)\big)$$
$$\big(Author(S, \; Lesson1) \rightarrow Autho^+ (S, \; Lesson1, Write)\big)$$
$$\big(Teacher(S, \; Lesson1) \rightarrow Autho^+ (S, \; Lesson1, Read)\big)$$
$$\big(Teacher(S, \; Lesson1) \rightarrow Autho^- (S, \; Lesson1, Write)\big)$$
$$\big(Student(S, \; Lesson1) \rightarrow Autho^+ (S, Lesson1, Read)\big)$$
$$\big(Autho^+ (S, \; Lesson1, A) Autho^- (S, \; Lesson1, A)\big) \rightarrow Autho(S, Lesson1, A)\big)$$

The first step towards the creation of our school e-Learning system is the standardization of key processes associated with the personalization of access to e-lessons.

The teachers create e-lessons in specialized SCORM-compliant and ontology-based development environment, then publish them in the education portal in a special Lesson-DB. Further to SCORM-metadata, we will use some additional specifications such as:

- **Info** – title of the lesson, school subject, author, etc. features that are supported by the SCORM-metadata;

- **Subdomain** – matrix with concepts that will be included in the lesson and the extent of their studying Subdomain(concept, m), where m=1,2,3 as: 1 – low level of studying (mandatory minimum, according to BES); 2 – good level и 3 – high level;

- **Num_Grade** (the grade, for which is intended the lesson) – an integer from 1 to 12;

- **Form_of_training** (form of training): 1 – regular training; 2 – self training;

- **Lesson_Status** (status of the lesson) – an integer between 1 and 4: 4-free for use by all users in this and other portals in DeLC-education network; 3-free for use only by students and teachers in the portal; 2-authorized use only for certain users; 1-unavailable for other users, except for the author; **Didactic_aims** (didactic aims, according to Bloom's taxonomy) – an integer between 1 and 5: 1 acquisition of new knowledge (level "knowledge" and "comprehension" in the Bloom's taxonomy); 2 actualization of old knowledge (level "comprehension", "application", and "analysis"); 3 exercise and improvement of knowledge (level "application", "analysis", and "synthesis"); 4-generalization (levels "analysis", "synthesis", and "evaluation"); and 5-exam (level "evaluation").

Therefore, any electronic lesson in the education portal is a vector with the above dimensions:

$$\text{Lesson}\big(\text{Info}(\text{ID, title, domain, author,...}), \text{Subdomain}(\text{concept}, m),$$
$$\text{Num\_Grade, Form\_of\_training, Lesson\_status, Didactic\_aims}\big)$$

For example, the lesson "Past imperfect tense of the verb", school subject "Bulgarian language", for 5th grade; author Sarafov, with concepts from matrix Subdomain, designed for regular students, free for use for all users in the education system, and is a lesson for new knowledge we get:

Lesson1 $\big($Info$\big($ID, Past imperfect tense of the verb,Bulgarian language, Saratov,...$\big)$,

Subdomain *5, 2, 4, 1$\big)$,

where Subdomain* is present with Table 1.

Table 1. Subdomain

Conceptions	Level of Studying
Verb	3
Person of the verb	2
Tense of the verb	2
Communication moment	3
Moment of action	3
Main orientation moment	2
Additional orientation moment	2

When the student requires launching of a lesson around a chosen theme, the system checks the availability of the appropriate e-lesson from the Lesson DB. Lessons that meet the initial user requirements are usually more than one, so the system should provide an appropriate mechanism for selecting the most appropriate among them. After a dialogue with the student, the personal agent defines his personal aims, preferences, etc., and transmits this vector to the system for choices. After the comparison with the vectors of the uploaded e-lessons in the Lesson DB, the e-Lessons with the highest level of similarity are extracted. The result will be a number of e-lessons andthe system should choose the most appropriate. This selection can be realized by the use of some intelligent algorithm (ex. CBR-approach).

The preferences and personal goals of each student can also formalize the policy which defines the sequence of actions in this training scenario. After the identification of the student in the training environment, based on the profile and persona-stereotypical information and a dialogue with his personal agent, the system receives the necessary initial values of the observed variables. After determining the initial state, the policy management can be transferred to a special Policy-Engine, which is part of the infrastructure of the run-time environment of the educational e-Learning portal. Initially, based on the dialogue with the student, the Policy of Preferences registers in the Policy-Engine and then starts the Mechanism for Selecting of Lesson that makes a request to the Lesson DB. After the selection a particular lesson, this e-lesson is filed to SCORM-Learning Management System for implementation. The scenario, which will run activities in the learning process, are described and formalized in the SCORM Sequence & Navigation-model and the corresponding parameterized template by which is created this lesson. Policy-Engine can continually modify policies according to the information coming from the behaviour of the learner.

The learning scenario may include mandatory implementation actions (e.g. solving tests). If a student fails to successfully complete these actions, the learning process falls in a critical condition and the Policy-Engine has to choose more appropriate lessons. In this case, the learning process is temporarily interrupted and the LMS restarts the training process with the new lesson.

The Policy of Preferences is expressed by the rules of condition-action types. Conditions present a number of behaviors that trigger certain actions. The formal semantics of the model is based on ITL as the rules are the following:

when B [increase | decrease] preference in Lesson [low | medium | high], where B is behaviour and Lesson is the e-lesson.

The degree of preference can be expressed as an integer. The larger number represents a higher degree of preference. It is initially assumed that the student doesn't have any preferences and all values are 0. We define the meaning of low, medium, and high level of preferences as 1, 2, and 3. We will look at an example of training with two lessons on the same learning material. The first lesson is more difficult and presents the studying concepts in a higher level than the second one. The student initially has not decided what his preferences are. In the Policy-Engine, there are defined policies, which specify that the lessons that guarantee more than 70% results in the final test are preferable than those that only guarantee between 50% to 70% and the lessons that ensure less than 50% are not preferred. We can express the policy with the following rules:

Score (Lesson1, Lesson2):

When (1: test_result >= 70%) increase preference in Lesson to high
When (2: 50%<=test_result <70%) decrease preference in Lesson to medium
When (3:test_result<50) decrease preference in Lesson to low

The Policy-Engine determines the information needed for the implementation according to these rules. LMS through the SCORM RTE[6] - and the mechanism of the target variables (objectives) determines the outcome of the student in solving the test. After each experience, the Policy-Engine checks the assumption as defined by the rules and determines whether they are appropriate. Let's assume that the student has an aim to study the learning material at a high level (3). The system starts Lesson1 and the results of the three consecutive attempts to resolve the final test are 55, 49%, and 60%. After the first attempt to solve the final test, the Policy-Engine activates the second rule because the result is between 50% and 70%. This determines the preference in 2. The next value is <50%. According to rule three, the Policy-Engine reduces the preference from 2 to 1. The last attempt to solve the test starts again with rule two and increases the preference from 1 to 2. This result is unsatisfactory for the personal aims of the student and as a result, the Policy-Engine defines the lesson as inappropriate. The learning process suspends the former lesson and continues with Lesson 2. The student's results from solving the final test for this lesson are 64%, 68%, and 72%. At the first attempt, the preferences rise to 2, the second is retained the same level, while the third attempt increases it to 3, which is quite satisfactory for the student's personal aims that the student has set.

The dynamic adaptive levels most directly correspond to the third type of interaction – Open Experience – because the communication is dynamic with continuous engagement between the system and student.

4. PERSONALIZATION AND USER MODELING

User modelling is an important feature of any e-learning system, to personalize and tailor the e-Learning to individual characteristics, knowledge, didactic aims, and the preferences of the students [19], [20], [21]. On the basis of the previous section, we can describe the adaptability of the system for e-Learning to knowledge and the preferences of students in elementary, static, and dynamic levels [22].

The Elementary Adaptive Level is guaranteed by the profile information about the student before starting his training process in the system. Based only on this adaptive level, the e-learning system offers only learning resources that are common to all students of the same grade and form of training.

The Static Adaptive Level is based on the model for selecting the most appropriate lesson from the Lessons DB as the student is joined to a particular persona in the stereotypical hierarchy. By personas, students with similar characteristics are presented in the e-learning system together. The lessons are prebuilt by the parameterization of the basic BPG-templates and models from the authors of e-learning content in a special authoring environment. These lessons are placed in a special repository – Lesson DB – and are described in metadata as described above.

The Dynamic Adaptive Level is implemented through the Policies of Preferences and Policy-Engine, which dynamically monitors the behaviours of student and his preferences with the relevant lesson in the actual learning process and can replace the current lesson with another that is more appropriate for the individual student.

In the adaptation process in terms of the user modelling, we will look at: the information athering about the learner, processing the information and its update, and finding and presenting the appropriate training resources for the considered student. The model describes the notion of the e-learning system for user knowledge, for his preferences, and aims. This model must be continuously updated according to the dynamic changes in the process of accumulation of knowledge about the particular student (Figure 4). The algorithm includes the following steps:

- Step 1. Filling the static profile information. According to the grade and form of training, the student is associated to any persona in a stereotypical hierarchy. The initial parameters are filled in interactive mode or the system gets the default values from the general stereotype model. Stereotyping and personas are used to transfer more general information about the group in the assumption of the individual user.

- Step 2. According to the persona, which is associated with the student, the system determines the common characteristics of the group and includes default values. Then, in the dialogue mode,the school subject, topic, and personal didactic aims of the student are determined. The rules are updated on the basis of collected information. The Policy-Engine launches the Searching Mechanism for the more appropriate lesson from the Lesson DB and submits it to the LMS for implementation.

- Step 3. The system manages individualized learning process. If there are any discrepancy found between personal aims,the knowledge of the student, and the rules, the Policy-Engine interrupts the current training scenario and restarts the Searching Mechanism for a new choice.

- Step 4. The system stores the new values of the parameters and change the rules by which Policy-Engine manages personal learning process of the individual student.

The information in this user model can be considered as information specific to the school subject domain and information that is independent of it. The first type includes the data connected with the Dynamic Adaptive Level as an evaluation of the student; his background knowledge and records of his behavior (number of passed lessons, number of errors during solving test, number of inappropriate lessons, etc.). The information that is not dependent on the some subject domain is related to the personal goals of the learner, with his motivation, experience, preferences, interests, and personal data such as name, years, type of training, etc.

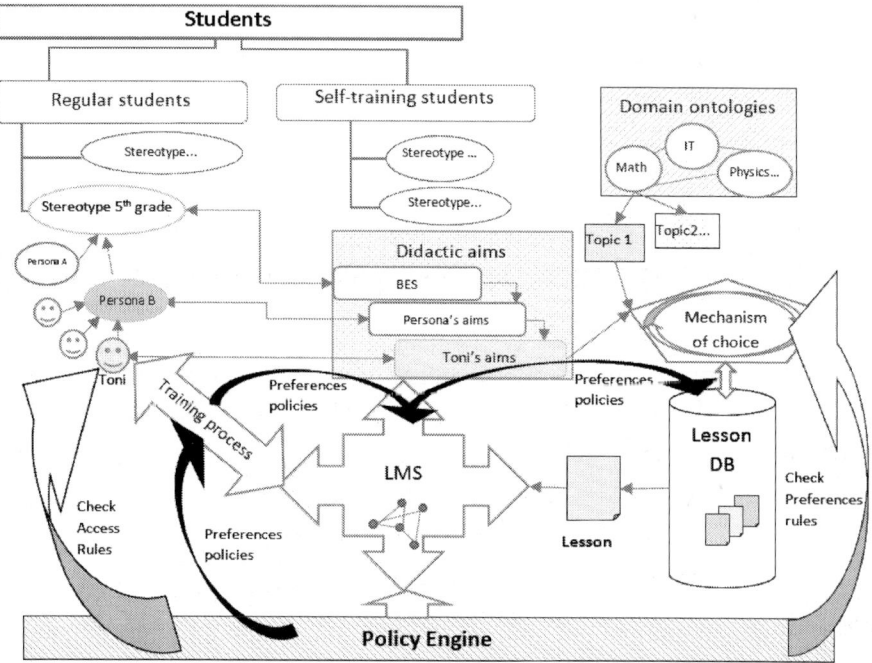

Figure 4. User modelling and personalization

The presented algorithm provides a continuous actualization of information. Such one is independent from the specific school subject domain and one that is domain-dependent. The model is continuously updated to correctly present the student in the e-Learning environment.

We created several versions of SCORM-based e-Learning portal of the secondary school "Hristo Smirnenski"-Brezovo, which is based on the conceptual framework of the system DeLC and supports SCORM RTE [23]. The latest version of the environment ensures the personalization in the elementary and static levels. We developed the mechanism of parameterization of the basic BPG-templates and models, and created an authoring tool for the designing and packaging of SCORM-based e-lessons.

Ontologies provide developers with predefined resources covering a specific school subject domain that can be used directly in the content. The establishment of educational environment is based on the adaptation of the corporate portal of the Delphi group. For the realization of the educational portal, we used the portal framework Liferay (http://liferay.com), which has implemented LMS of SCORM RTE [24]. There are many services implemented into the portal that supports the training process in different subjects and raises the level of interactivity in learning.

5. CONCLUSIONS AND FUTURE WORK

The proposed user model allows to increase the level of personalization in the e-Learning system. This is essential for learners from all forms of training, but is particularly important for students using the distance form of training, pupils with special educational needs, and disabled children. The implementation at the elementary level of the model is provided by the Autho - rules, which depend on stereotyped groups and personas with their access rights to portal resources. Users could be students, teachers, parents, authors of learning content, and so on. If they are students, access must be allowed to educational materials for the appropriate grade, form of training, and so on. If they are teachers, according to their stereotypical information, the mechanism provides them an access to learning resources and services related to their school subjects. If they are authors of educational content, they are allowed access to the Lesson DB for editing and adding of e-Lessons. If they are parents, they are allowed an access to information about their children. Different scenarios for access formalize a sequence of different rules for each group of users managed from the Policy-Engine. The second level of user-modelling is realized through the model for selecting the most appropriate lesson from the repository of lessons – lesson DB and their meta-description by the lesson-vector. The Dynamic Adaptive Level is implemented through the set of Policies of Preferences. The Policy-Engine monitors the behaviour of students and their preference at the relevant lesson and can dynamically replace the current lesson with a more appropriate one.

Based on the MOOC Integrated Learning Environment (http://ilde.upf.edu/handson3), the authors of e-Learning resources successively pass through several steps – the definition of didactic aims, analysis of educational content, creation of personas, designing of learning scenario, creation of e-Lesson in specialized SCORM authoring environment, share a draft version of the created e-lesson for evaluation, and heuristic analysis from other pedagogical specialists. At each step, the authors directly share with their counterparts in the integrated environment. After the completion of the first cycle and depending on the evaluation, authors can then correct their lessons and again pass through the step-by-step cycle or publish the e-lesson in the Lesson DB of education portal. This process is cyclical and leads to the continuous improvement and refinement of the developed learning resources. It meets specified infigure 1 workflow. The published lessons are created in dialogue with educational specialists and tested directly in the real learning environment. This largely ensures the sustainability of the model and overcomes some of the problems in the development of e-Learning.The report is part of the work on the project IT 15- FMIIT-004 "Research in the field of innovative ICT business orientation and training" to fund "Scientific Researches" of Plovdiv University "Paisii Hilendarski".

NOTES

[1] - DeLC- Distributed eLearning Centre

[2] - SCORM- Sharable Content Object Reference Model

[3] - SCO- Sharable Content Object

[4] - CAM-SCORM Content Aggregation Model

[5] - S&N Model- SCORM Sequence and Navigation Model

[6] - RTE- Run-Time Environment

REFERENCES

1. Carliner, S., Patti Shank, editors, The e-Learning handbook: past promises, present challenges. 1st ed. USA: Pfeiffer; 2008. 560 p., ISBN: 978-0-7879-7831-0

2. Allen, I. Elaine; Seaman, Jeff. Entering the Mainstream: The Quality and Extent of Online Education in the United States, 2003 and 2004. 1st ed. USA: Sloan Consortium (Sloan-C); 2004. 27 p., ISBN:0-9677741-8-7

3. Abrami, R., Barrett, H., Directions for research and development on electronic portfolios. Canadian Journal of Learning and Technology. 2005; 31(3):1-15, ISSN 1499-6685

4. Drucker, P., Need to Know: Integrating e-Learning with High Velocity Value Chains. 1st ed. USA: Delphi Group; 2000. 12 p., Delphi Group White Paper

5. Stoyanov, S., I. Ganchev, I. Popchev, M. O'Droma, From CBT to e-Learning. Information Technologies and Control. 2005;3(4):2-10, ISSN 1312-2622.

6. Stoyanovich, L.,Staab S., Studer R.,. e- Learning, based on the Semantic Web. In: WebNet2001-World Conference on the WWW and Internet; 23-27.10.2001; Orlando, Florida. USA:2001. p. 23-27, ISBN 1-880094-46-0.

7. Shedroff, N., A unified field theory of Design. In: Jacobsen R., editor. Information Interaction Design. Cambridge: MIT Press; 1999. p. 267-292.

8. Glushkova, T., Adaptive Model for E-Learning in Secondary School. In: Elvis Pontes, editor. E-Learning – Long-Distance and Lifelong Perspectives. 1st ed. Croatia: InTech; 2012. p. 3-22, ISBN 978-953-51-0250-2. DOI: 10.5772/29342

9. Glushkova, T., Adaptive environment for e-Learning in secondary schools [PhD thesis]. Plovdiv University: Plovdiv University; 2011. 220 p. Available from: http://procedures.uni-plovdiv.bg/docs/ procedure/163/1820394915693780406.pdf

10. Lene, N., Personas. In: Rikke Friis, editor. The Encyclopedia of Human-Computer Interaction. 2nd ed. Denmark: The Interaction Design

Foundation; 2014. https://www.interaction-design.org/ encyclopedia/ personas.html.

11. Mor, Y., Mogilevsky, O., Learning Design Studio: Educational Practice as Design Inquiry of Learning. In: EC-TEL 20013; 17-21 September 2013; Paphos. Berlin: Springer; 2013. p. 233-245.

12. Glushkova, T., Trendafilova, M.,Uzunova, N., Application of SCORM standard for e-Learning in secondary school. In: Informatics in the Scientific Knowlege- ISK'2006; 18-22 June 2006; Varna. Varna: VFU; 2006. p. 205-216, ISSN 1313-4345, ISBN – 10:954-715-303-X, ISBN-13:978-954-715-303-5.

13. Stoyanov, S., D. Mitev, I. Minov, T. Glushkova. E-Learning Development Environment for Software Engineering Selbo 2. In: 19th Database and Expert Systems Application (DEXA 2008); 1-5 September 2008; Turin. Springer: 2008. p. 100-104, ISBN: 978-3-540-85653-5. DOI: 10.1109/DEXA.2008.89

14. Bloom, B., Taxonomy of Educational Objectives Book 1: Cognitive Domain. 2nd ed. Addison Wesley Publishing Company; 1984 (first published June 1956). 204 p., ISBN13: 9780582280106

15. Learning System Architecture Lab. SCORM Best Practices Guide for Content Developers. 1st ed. Pittsburgh, USA: Carnegie Mellon University; 2003. 80 p.

16. Mitev, D., Popchev I. Intelligent agents and sevices in eLearning development environment Selbo 2. In: ISK'2008; 26-28 June 2008; Varna. Bulgaria: ISK; 2008. p. 275-284. DOI: ISBN-13:978-954-715-303-5

17. Sloman, M.. Policy driven management for distributed systems. Journal of network and Systems Management. 1994; 2(3):333-360, ISSN 1064-75-70 (Print), 1573-7705(online).

18. Moskowski, B. Reasoning about Digital Circuits [Dissertation]. Department of computer science: Stanford University Stanford, CA, USA; 1983. 148 p. Available from: http://dl.acm.org/citation. cfm?id=911281 DOI: AAI8329756

19. Kobsa, A. User modelling and user-Adapted Interaction. Springer Netherlands. 2004; 14(5):469-475, ISSN: 0924-1868(paper) 1573-1391(online). DOI: 10.1007/s11257-005-2618-3

20. Glushkova,T., User modelling of distributed e-Learning system for secondary school. In: DIDMATHTECH; 2006; Komarno. Slovakia: DIDMATHTECH; 2006. p. 117-123. ISBN: 978-80-89234-23-3

21. Brusilovsky, P. Adaptive hypermedia. User Modelling and User Adapted Interaction, Ten Year Anniversary Issue. 2001; 11(1-2):87-110, ISSN: 0924-1868.

22. Glushkova, T., Stoyanova, A., Interaction and adaptation to the specificity of the subject domains in the system for e-Learning and Distance training DeLC. In: Informatics in the Scientific Knowlege; 17-20 June 2008; Varna. Bulgaria: ISK; 2008. p. 295-307. ISSN: 1313-4345, ISBN: 13:978-954-715-303-526-28

23. Glushkova, T, Stoyanov, S., Trendafilova, M, Cholakov, G., Adaptation of DeLC system for e_learning in Secondary school. In: CompSysTech'2005; 16-18 June 2005; Varna. Bulgaria: CompSysTech; 2005. p. IV.15.1-15.6. ISBN: 954-9641-38-4

24. Glushkova, T., E-Learning environment for supporting of secodary school education. Cybernetics and Information Technologies. 2007; 7(3):89-106, ISSN: 1311-9702.

Index

3D Interactions, 171

A

adaptability, 133, 216, 223, 275, 277, 279, 294

Adoption, 59, 60, 83, 85, 272

Assessment guidelines, 198

Asynchronous, 169, 247, 248, 269

C

Canvas, 111, 117, 118, 121, 122, 124

D

Digital Textbook, 60

disability, 216, 233, 235

Disease Management, 127, 128, 140, 141, 142, 144, 145

distance education, 24, 26, 27, 36, 42, 56, 57, 169, 170, 215, 243

distance language learning, 24, 27, 42

DM Nurses, 130

documentation, 1, 2, 116

E

Education Program, 128, 143, 201

e-inclusion, 216

E-Learning, 9

E-LIS, 1, 2, 13, 15, 16, 17, 18

eTeaching, 23, 24, 25, 33, 34, 46, 53, 54

eTeaching and eLearning, 24

I

Information Communication Technology (ICT), 247, 248

Instructional websites, 198

Interaction, 2, 14, 19, 100, 147, 148, 154, 168, 273, 275, 276, 299, 301

intermediate concepts, 88, 89, 90, 103

Internet, 125, 181, 198, 201, 202, 204, 205, 208, 209, 211, 212, 218, 224, 225, 242, 243, 244, 245, 247, 248, 251, 262, 271, 299

Internet server end, 198

ITL, 273, 279, 287, 292

K

knowledge building, 147

L

learner traits, 46

learning objects, 63, 195, 216, 217, 218, 223, 230, 231, 233, 241, 242

Learning process mechanism, 147

LMS, 111, 115, 117, 157, 193, 217, 218, 219, 220, 222, 225, 228, 239, 241, 252, 281, 283, 286, 287, 292, 293, 295, 297

M

Moodle, 111, 115, 117, 118, 122, 124, 244

moral reasoning competency development, 88

N

National University of Mongolia, 111, 112

Nyust Students Group, 209

O

online instructor training, 24

Open Textbook, 60, 82

P

Personalization, 273, 276, 294

pharmacy ethics education, 88, 93

Ping Pong, 1, 2, 3, 4, 6, 7, 8

Q

Qualitative Weight and Sum Approach, 111, 113, 116

S

SCORM, 120, 223, 232, 238, 244, 273, 276, 281, 283, 284, 285, 287, 289, 290, 292, 293, 296, 298, 300

Self-regulated learning, 27

self-regulation, 23, 24, 30, 32, 33, 36, 46, 51, 57

services, 1, 2, 5, 13, 16, 17, 18, 21, 84, 112, 122, 129, 176, 177, 179, 180, 181, 191, 194, 195, 222, 235, 274, 281, 297

Student Preferences, 59, 60

T

technology enhanced learning, 57, 88

Telenursing, 128, 138, 144

The University of the South Pacific, 60

Theoretical integration, 32

U

user-centered, 2, 20

W

Web-Based, 56, 143, 144, 168, 248, 255, 263, 271, 272